Encyclopaedia of

Sustainable Water Resource Management

Encyclopaedia of

Sustainable Water Resource Management

Volume 5

Editor:

R.N. Sahay

AGROTECH PRESS

JAIPUR ▪ NEW DELHI

AGROTECH PRESS

Regd. Off. "Bony Residency", Gate No. 2
Opp. Tilak Public School, Vishweswariya Nagar,
Gopalpura Road, Jaipur - 302018, India
Ph: 0141 - 2761280, Telefax: 2761381
E-mail : agrotechpress@gmail.com

Delhi Office: 102, 1st Floor, Satyam House,
4327/3, Ansari Road, Darya Ganj,
New Delhi - 110002, India
Ph: 011-45652440
visit us: www.agrotechpress.com

Distributed By
KEYA INTERNATIONAL PRESS LLP
E-mail: keyainternationalpressllp@gmail.com

Published in India, 2025

ISBN : 9789384568269 (5 Vols. Set)

Typeset by Thomson Press

Printed and bound in India by Thomson Press Ltd.,
New Delhi

Contents

Preface

Human development and population growth exert many and diverse pressures on the quality and quantity of water resources and access to them. During the past few decades the stress on the availability of water is growing and its quality is deteriorating. The water resources sector is presently encountering a series of interlinked issues which if not dealt in totality may prove to have serious ramifications in the long run. Population explosion, unpredictable rainfall pattern, natural calamities and climate change further affect water resources. It is common knowledge that there is no life without water; the reference is not just water but safe water.

There is growing recognition of the need for increased access to drinkable water across the world and for water treatment approaches that improve the quality of the delivered water and re-establish a balance between human and natural systems. In the current paradigm, however, large-scale, centralized water treatment systems are slow to accommodate changes in supply or demand, and impede innovations that could address these issues because there are too many perceived risks—financial, technical, system, and organizational—associated with introducing change into these systems. Localized, networked water treatment systems improve access to potable water, encourage the development and diffusion of innovations through reduced financial and technical risks, lower the potential of total system failure, and provide easier trial and replacement of specific innovations and greater organizational capacity. These systems can also be tailored to minimize environmental damage and could enhance or encourage the deployment of in situ power generation. There are an encouraging number of innovative technologies, systems, components, processes, and management approaches emerging for localized water treatment, including new filtration and disinfectant technologies, treatment systems, business models, and approaches to effectively managing regional water resources. But to provide communities with drinking water will require a paradigm shift by the government agencies around the world responsible for public health, safety, and welfare and for protecting natural ecosystems. Only by changing our methods of water-treatment system procurement and management can we deliver equitable access to potable water. There are an encouraging number of innovative technologies, systems, components, processes, and management approaches emerging for localized water treatment, including new filtration and disinfectant technologies, treatment systems, business models, and approaches to effectively managing regional water resources. But to provide communities with drinking water will require a paradigm shift by the government agencies around the world responsible for public health, safety, and welfare and for protecting natural ecosystems. Only by changing our methods of water-treatment system procurement and management can we deliver equitable access to potable water. Traditional water treatment approaches are not effectively meeting these growing needs. Fortunately, a new paradigm is emerging: an increase in localized, networked water treatment systems and a parallel rise in on-site energy generation systems, as both infrastructures move from large, centralized production to smaller, interconnected units that can respond to local conditions and demands. The energy sector has seen a surge in innovations and new market opportunities for a host of entrants and established companies, a pattern that can be expected to occur in water treatment as well. In the same way that clean energy is expected to produce new jobs, businesses, and markets, localized water treatment can provide the same opportunities in economic development, quality of life, and environmental regeneration. In some areas, there have been exciting combinations of water treatment and energy generation on the local scale. The shift to localized water treatment networks can provide significant opportunities for further innovation developments. For instance, each water source (e.g., surface, ground, rain, or waste water) has its own specific characteristics in terms of turbidity, contaminants, dissolved chemicals, acidity, etc., and the innovations will need to specifically address these characteristics to improve quality, taste, and other attributes.

This book highlights the benefits of water re-use; water conservation; role of public participation and community awareness; water-resources development and capacity building for sustainable water management. This work should, therefore, be a welcome addition to all libraries interested in the history and development of water management in India.

—Editor

1

Waste-water Treatment Technologies

Activated Sludge

Activated sludge is a process for treating sewage and industrial wastewaters using air and a biological floc composed of bacteria and protozoans.

Purpose

In a sewage (or industrial wastewater) treatment plant, the activated sludge process can be used for one or several of the following purposes:

- oxidising carbonaceous matter: biological matter.
- oxidising nitrogeneous matter: mainly ammonium and nitrogen in biological materials.
- removing phosphate.
- driving off entrained gases carbon dioxide, ammonia, nitrogen, etc.
- generating a biological floc that is easy to settle.
- generating a liquor that is low in dissolved or suspended material.

The Process

The process involves air or oxygen being introduced into a mixture of primary treated or screened sewage or industrial wastewater (called wastewater from now on) combined with organisms to develop a biological floc which reduces the organic content of the sewage. This material, which in healthy sludge is a brown floc, is largely composed of saprotrophic bacteria but also has an important protozoan flora mainly composed of amoebae, Spirotrichs, Peritrichs including Vorticellids and a range of other filter feeding species.

Other important constituents include motile and sedentary Rotifers. In poorly managed activated sludge, a range of mucilaginous filamentous bacteria can develop including *Sphaerotilus natans* which produces a sludge that is difficult to settle and can result in the sludge blanket decanting over the weirs in the settlement tank to severely contaminate the final effluent quality. This material is often described as sewage fungus but true fungal communities are relatively uncommon.

The combination of wastewater and biological mass is commonly known as mixed liquor. In all activated sludge plants, once the wastewater has received sufficient treatment, excess mixed liquor is discharged into settling tanks and the treated supernatant is run off to undergo further treatment before discharge. Part of the settled material, the sludge, is returned to the head of the aeration system to re-seed the new wastewater entering the tank. This fraction of the floc is called return activated sludge (R.A.S.). Excess sludge is called surplus activated sludge (S.A.S.) or waste activated sludge (W.A.S). S.A.S is removed from the treatment process to keep the ratio of biomass to food supplied in the wastewater in balance. S.A.S is stored in sludge tanks and is further treated by digestion, either under anaerobic or aerobic conditions prior to disposal.

Many sewage treatment plants use axial flow pumps to transfer nitrified mixed liquor from the aeration zone to the anoxic zone for denitrification. These pumps are often referred to as internal mixed liquor recycle pumps (IMLR pumps). The raw sewage, the RAS, and the nitrified mixed liquor are mixed by submersible mixers in the anoxic zones in order to achieve denitrification. Activated sludge is also the name given to the active biological material produced by activated sludge plants.

History

The activated sludge process was discovered in 1913 in the UK by two engineers, Edward Ardern and W.T. Lockett, conducting research for the Manchester Corporation Rivers Department at Davyhulme Sewage Works. Dr G Fowler, co-founder of the activated sludge process should have the credit for originating the process even though Ardern and Lockett did much to develop it. All three of these men are better described as chemists than engineers. Experiments on treating sewage in a draw-and-fill reactor (the precursor to today's sequencing batch reactor) produced a highly treated effluent. Believing that the sludge had been activated (in a similar manner to activated carbon) the process was named *activated sludge*. Not until much later was it realised that what had actually occurred was a means to concentrate biological organisms, decoupling the liquid retention time (ideally, low, for a compact treatment system) from the solids retention time (ideally, fairly high, for an effluent low in BOD_5 and ammonia.)

Arrangement

The general arrangement of an activated sludge process for removing carbonaceous pollution includes the following items:

- Aeration tank where air (or oxygen) is injected in the mixed liquor.
- Settling tank (usually referred to as "final clarifier" or "secondary settling tank") to allow the biological flocs to settle, thus separating the biological sludge from the clear treated water.

Treatment of nitrogenous matter or phosphate involves additional steps where the mixed liquor is left in anoxic condition (meaning that there is no residual dissolved oxygen).

Types of Plants

Package Plants: There are a wide range of other types of plants, often serving small communities or industrial plants that may use hybrid treatment processes often involving the use of aerobic sludge to treat the incoming sewage. In such plants the primary settlement stage of treatment may be omitted. In these plants, a biotic floc is created which provides the required substrate.

Package plants are commonly variants of extended aeration, to promote the 'fit & forget' approach required for small communities without dedicated operational staff. There are various standards to assist with their design.

Oxidation Ditch

In some areas, where more land is available, sewage is treated in large round or oval ditches with one or more horizontal aerators typically called brush or disc aerators which drive the mixed liquor around the ditch and provide aeration. These are oxidation ditches, often referred to by manufacturer's trade names such as Pasveer, Orbal, or Carrousel. They have the advantage that they are relatively easy to maintain and are resilient to shock loads that often occur in smaller communities (i.e. at breakfast time and in the evening).

Oxidation ditches are installed commonly as 'fit & forget' technology, with typical design parameters of a hydraulic retention time of 24 - 48 hours, and a sludge age of 12 - 20 days. This compares with nitrifying activated sludge plants having a retention time of 8 hours, and a sludge age of 8 - 12 days.

Deep Shaft

Where land is in short supply sewage may be treated by injection of oxygen into a pressured return sludge stream which is injected into the base of a deep columnar tank buried in the ground. Such shafts may be up to 100 metres deep and are filled with sewage liquor. As the sewage rises the oxygen forced into solution by the pressure at the base of the shaft breaks out as molecular oxygen providing a highly efficient source of oxygen for the activated sludge biota. The rising oxygen and injected return sludge provide the physical mechanism for mixing of the sewage and sludge. Mixed sludge and sewage is decanted at the surface and separated into supernatant and sludge components. The efficiency of deep shaft treatment can be high.

Surface aerators are commonly quoted as having an aeration efficiency of 0.5 - 1.5 kg O_2/kWh, diffused aeration as 1.5 - 2.5 kg O_2/KWh. Deep Shaft claims 5 - 8 kg O_2/kWh. However, the costs of construction are high. Deep Shaft has seen greatest uptake in Japan, because of the land area issues. Deep Shaft was developed by ICI, as a spin-off from their Pruteen process.

In the UK it is found at three sites: Tilbury, Anglian water, treating a wastewater with a high industrial contribution; Southport, United Utilities, because of land space issues; and Billingham, ICI, again treating industrial effluent, and built (after the Tilbury shafts) by ICI to help the agent sell more.

DeepShaft is a patented, licensed, process. The licensee has changed several times and, currently (2007), it is Aker Kvaerner Engineering Services.

Surface-aerated Basins/Lagoons

Most biological oxidation processes for treating industrial wastewaters have in common the use of oxygen (or air) and microbial action. Surface-aerated basins achieve 80 to 90 per cent removal of BOD with retention times of 1 to 10 days. The basins may range in depth from 1.5 to 5.0 metres and utilise motor-driven aerators floating on the surface of the wastewater.

In an aerated basin system, the aerators provide two functions: they transfer air into the basins required by the biological oxidation reactions, and they provide the mixing required for dispersing the air and for contacting the reactants (that is, oxygen, wastewater and microbes). Typically, the floating surface aerators are rated to deliver the amount of air equivalent to 1.8 to 2.7 kg O_2/kWh. However, they do not provide as good mixing as is normally achieved in activated sludge systems and therefore aerated basins do not achieve the same performance level as activated sludge units.

Biological oxidation processes are sensitive to temperature and, between 0°C and 40°C, the rate of biological reactions increase with temperature. Most surface aerated vessels operate at between 4°C and 32°C.

Aeration Methods

Diffused Aeration

Sewage liquor is run into deep tanks with diffuser grid aeration systems that are attached to the floor. These are like the diffused airstone used in tropical fish tanks but on a much larger scale. Air is pumped through the blocks and the curtain of bubbles formed both oxygenates the liquor and also provide the necessary mixing action. Where capacity is limited or the sewage is unusually strong or difficult to treat, oxygen may be used instead of air. Typically, the air is generated by some type of blower or compressor.

Jet Aerators

A jet aerator works through aspirating technology and thus does not require any external air source (i.e. compressor), except for the surrounding atmosphere. Jet aerators can be installed either as submersible units or piped through the tank wall using an external dry-installed chopper pump to feed the aspirating ejector(s).

Surface Aerators (Cones)

Vertically mounted tubes of up to 1 metre diameter extending from just above the base of a deep concrete tank to just below the surface of the sewage liquor. A typical shaft might be 10 metres high. At the surface end the tube is formed into a cone with helical vanes attached to the inner surface.

When the tube is rotated, the vanes spin liquor up and out of the cones drawing new sewage liquor from the base of the tank. In many works each cone is located in a separate cell that can be isolated from the remaining cells if required for maintenance. Some works may have two cones to a cell and some large works may have 4 cones per cell.

Advanced Oxidation Process

Advanced oxidation processes (abbreviation: AOPs) refers to a set of chemical treatment procedures designed to remove organic and inorganic materials in waste water by oxidation. One such type of process is called in situ chemical oxidation. Contaminants are oxidised by four different reagents: ozone, hydrogen peroxide, oxygen, and air, in precise, pre-programmed dosages, sequences, and combinations. These procedures may also be combined with UV irradiation and specific catalysts.

This results in the development of hydroxyl radicals. A well known example of AOP is the use of Fenton's reagent. The AOP procedure is particularly useful for cleaning biologically toxic or non-degradable materials such as aromatics, pesticides, petroleum constituents, and volatile organic compounds in waste water .

The contaminant materials are converted to a large extent into stable inorganic compounds such as water, carbon dioxide and salts, i.e. they undergo mineralisation. A goal of the waste water purification by means of AOP procedures is the reduction of the chemical contaminants and the toxicity to such an extent that the cleaned waste water may be reintroduced into receiving streams or, at least, into a conventional sewage treatment.

Aerated Lagoon

An aerated lagoon or aerated basin is a holding and/or treatment pond provided with artificial aeration to promote the biological oxidation of wastewaters. There are many other biological processes for treatment of wastewaters, for example activated sludge, trickling filters, rotating biological contactors and biofilters. They all have in common the use of oxygen (or air) and microbial action to biotreat the pollutants in wastewaters.

Types of Aerated Lagoons or Basins

- Suspension mixed lagoons, where there is less energy provided by the aeration equipment to keep the sludge in suspension.
- Facultative lagoons, where there is insufficient energy provided by the aeration equipment to keep the sludge in suspension and solids settle to the lagoon floor. The biodegradable solids in the settled sludge then degrade anaerobically.

Suspension Mixed Lagoons

Suspension mixed lagoons flow through activated sludge systems where the effluent has the same composition as the mixed liquor in the lagoon. Typically the sludge will have a residence time or sludge age of 1 to 5 days. This means that the chemical oxygen demand (COD) removed is relatively little and the effluent is therefore unacceptable for discharge into receiving waters.

The objective of the lagoon is therefore to act as a biologically assisted flocculator which converts the soluble biodegradable organics in the influent to a biomass which is able to settle as a sludge. Usually the effluent is then put in a second pond where the sludge can settle. The effluent can then be removed from the top with a low COD, while the sludge accumulates on the floor and undergoes anaerobic stabilisation. .

Methods of Aerating Lagoons or Basins

There are many methods for aerating a lagoon or basin:

- Motor-driven floating surface aerators
- Motor-driven submerged aerators
- Motor-driven fixed-in-place surface aerators
- Injection of compressed air through submerged diffusers.

Floating Surface Aerators

Ponds or basins using floating surface aerators achieve 80 to 90 per cent removal of BOD with retention times of 1 to 10 days. The ponds or basins may range in depth from 1.5 to 5.0 metres.

In a surface-aerated system, the aerators provide two functions: they transfer air into the basins required by the biological oxidation reactions, and they provide the mixing required for dispersing the air and for contacting the reactants (that is, oxygen, wastewater and microbes). Typically, the floating surface aerators are rated to deliver the amount of air equivalent to 1.8 to 2.7 kg O_2/kWh. However, they do not provide as good mixing as is normally achieved in activated sludge systems and therefore aerated basins do not achieve the same performance level as activated sludge units.

Biological oxidation processes are sensitive to temperature and, between 0°C and 40°C, the rate of biological reactions increase with temperature. Most surface aerated vessels operate at between 4°C and 32°C.

Submerged Diffused Aeration

Submerged diffused air is essentially a form of a diffuser grid inside a lagoon. There are two main types of submerged diffused aeration systems for lagoon applications: floating lateral and submerged lateral. Both these systems utilise fine or medium bubble diffusers to provide aeration and mixing to the process water. The diffusers can be suspended slightly above the lagoon floor or may rest on the bottom. Flexible airline or weighted air hose supplies air to the diffuser unit from the air lateral (either floating or submerged).

Aerobic Granular Reactor

Aerobic granular reactor is a type of waste treatment facility.

Overview

Wastewater treatment plants (WWTP) based on activated sludge often cover large surface areas, mainly caused by the large settling tanks. To be able to build compact WWTP's biomass can be grown as biofilms on a carrier material or as fast settling aerobic granular sludge without a carrier. Recent research showed the advantages of a discontinuously fed system, in which it is possible to grow stable granulated sludge under aerobic circumstances.

Simultaneous Chemical oxygen demand, Nitrogen and Phosphorus removal can be easily integrated in a discontinuous fed system. Because of the high settling capacity of the granules, the use of traditional settler is unnecessary and therefore the installation can be built very compact, needing only 20 per cent of the surface area of conventional activated sludge systems.

Aerobic Treatment System

An aerobic treatment system or ATS, often called (incorrectly) an aerobic septic system is a small scale sewage treatment system similar to a septic tank system, but which uses an aerobic process for digestion rather than just the anaerobic process used in septic systems. These systems are commonly found in rural areas where public sewers are not available, and may be used for a single residence or for a small group of homes. Unlike the traditional septic system, the aerobic treatment system produces a high quality secondary effluent, which can be sterilised and used for surface irrigation. This allows much greater flexibility in the placement of the leach field, as well as cutting the required size of the leach field by as much as half.

Process

The ATS process generally consists of the following phases:

- Pre-treatment stage to remove large solids and other undesirable substances from the wastewater; this stage acts much like a septic system, and an ATS may be added to an existing septic tank to further process the primary effluent.
- Aeration stage, where the aerobic bacteria digest the biological wastes in the wastewater.
- Settling stage to allow any undigested solids to settle. This forms a sludge which must be periodically removed from the system.
- Disinfecting stage, where chlorine or similar disinfectant is mixed with the water, to produce an antiseptic output.

The disinfecting stage is optional, and is used where a sterile effluent is required, such as cases where the effluent is distributed above ground. The disinfectant typically used is tablets of calcium hypochlorite, which are specially made for waste treatment systems. Unlike the chlorine tablets used in swimming pools, which is stabilised for resistance to breakdown in ultraviolet light, the tablets used in waste treatment systems is intended to break down quickly in sunlight. Stabilised forms of chlorine will persist after the effluent is dispersed, and can kill off plants in the leach field. Since the ATS contains a living ecosystem of microbes to digest the waste products in the water, excessive amounts of items such as bleach or antibiotics can damage the ATS environment and reduce treatment effectiveness. Non-digestible items should also be avoided, as they will build up in the system and require more frequent sludge removal.

Types of Aerobic Treatment Systems

Small scale aerobic systems generally use one of two designs, fixed-film systems, or continuous flow, suspended growth aerobic systems (CFSGAS). The pre-treatment and effluent handling are similar for both types of systems, and the difference lies in the aeration stage.

Fixed Film Systems

Fixed film systems use a porous medium which provides a bed to support the biomass film that digests the waste material in the wastewater. Designs for fixed film systems vary widely, but fall into two basic categories (though some systems may combine both methods). The first is a system where the media is moved relative to the wastewater, alternately immersing the film and exposing it to air, while the second uses a stationary media, and varies the wastewater flow so the film is alternately submerged and exposed to air. In both cases, the biomass must be exposed to both wastewater and air for the aerobic digestion to occur.

The film itself may be made of any suitable porous material, such as formed plastic or peat moss. Simple systems use stationary media, and rely on intermittent, gravity driven wastewater flow to provide periodic exposure to air and wastewater. A common moving media system is the rotating biological contactor (RBC), which uses disks rotating slowly on a horizontal shaft. Approximately 40 percent of the disks are submerged at any given time, and the shaft rotates at a rate of one or two revolutions per minute.

Continuous Flow, Suspended Growth Aerobic Systems

CFSGAS systems, as the name imply, are designed to handle continuous flow, and do not provide a bed for a bacterial film, relying rather on bacteria suspended in the wastewater. The suspension and aeration are typically provided by an air pump, which pumps air through the aeration chamber, providing a constant stirring of the wastewater in addition to the oxygenation. A medium to promote fixed film bacterial growth may be added to some systems designed to handle higher than normal levels of biomass in the wastewater.

Retrofit or Portable Aerobic Systems

Another increasingly common use of aerobic treatment is for the remediation of failing or failed anaerobic septic systems, by retrofitting an existing system with an aerobic feature. This class of product, known as aerobic remediation, is designed to remediate biologically failed and failing anaerobic distribution systems by significantly reducing the BOD5 and TSS of the effluent. The reduction of the BOD5 and TSS reverses the developed bio-mat. Further, effluent with high dissolved oxygen and aerobic bacteria flow to the distribution component and digest the bio-mat.

Composting Toilets

Composting toilets are designed to treat only toilet waste, rather than general residential waste water, and are typically used with water-free toilets rather than the flush toilets associated with the above types of aerobic treatment systems. These systems treat the waste as a moist solid, rather than in liquid suspension, and therefore separate urine from feces during treatment to maintain the correct moisture content in the system. An example of a composting toilet is the clivus multrum (Latin for 'inclined chamber'), which consists of an inclined chamber that separates urine and feces and a fan to provide positive ventilation and prevent odours from escaping through the toilet. Within the chamber, the urine and feces are independently broken down not only by aerobic bacteria, but also by fungi, arthropods, and earthworms.

Treatment times are very long, with a minimum time between removals of solid waste of a year; during treatment the volume of the solid waste is decreased by 90 per cent, with most being converted into water vapour and carbon dioxide. Pathogens are eliminated from the waste by the long durations in inhospitable conditions in the treatment chamber.

Comparison to Traditional Septic Systems

The aeration stage and the disinfecting stage are the primary differences from a traditional septic system; in fact, an aerobic treatment system can be used as a secondary treatment for septic tank effluent. These stages increase the initial cost of the aerobic system, and also the maintenance requirements over the passive septic system. Unlike many other biofilters, aerobic treatment systems require a constant supply of electricity to drive the air pump increasing overall system costs.

The disinfectant tablets must be periodically replaced, as well as the electrical components (air compressor) and mechanical components (air diffusers). On the positive side, an aerobic system produces a higher quality effluent than a septic tank, and thus the leach field can be smaller than that of a conventional septic system, and the output can be discharged in areas too environmentally sensitive for septic system output. Some aerobic systems recycle the effluent through a sprinkler system, using it to water the lawn where regulations approve.

Effluent Quality

Since the effluent from an ATS is often discharged onto the surface of the leach field, the quality is very important. A typical ATS will, when operating correctly, produce an effluent with less than 30 mg/litre biochemical oxygen demand, 25 mg/litre total suspended solids, and 10,000 cfu/mL fecal coliform bacteria. This is clean enough that it cannot support a biomat or "slime" layer like a septic tank.

ATS effluent is relatively odourless; a properly operating system will produce effluent that smells musty, but not like sewage. Aerobic treatment is so effective at reducing odours, that it is the preferred method for reducing odour from manure produced by farms.

Anaerobic Clarigester

The anaerobic clarigester is a form of anaerobic digester. It is regarded as being the ancestor of the UASB anaerobic digester. A clarigester treats dilute biodegradable feedstocks and separates out solid and hydraulic (liquid) retention times. A diagram comparing the UASB, anaerobic clarigester and anaerobic contact processes can be found here.

Anaerobic Filter

In sewage treatment systems, an anaerobic filter (AF) is a form of anaerobic digester. The digestion tank contains a filter medium where anaerobic microbial populations — organisms that live in the absence of oxygen — can establish themselves. Such filters are commonly employed in the treatment of waste water. These reactors are gaining in popularity versus more established aerobic waste-water treatment systems because they produce a less solid residue than do other types of filter.

API Oil-water Separator

An API oil-water separator is a device designed to separate gross amounts of oil and suspended solids from the wastewater effluents of oil refineries, petrochemical plants, chemical plants, natural gas processing plants and other industrial sources. The name is derived from the fact that such separators are designed according to standards published by the American Petroleum Institute (API).

Description of the Design and Operation

The API separator is a gravity separation device designed by using Stokes Law to define the rise velocity of oil droplets based on their density and size. The design of the separator is based on the specific gravity difference between the oil and the wastewater because that difference is much smaller than the specific gravity difference between the suspended solids and water.

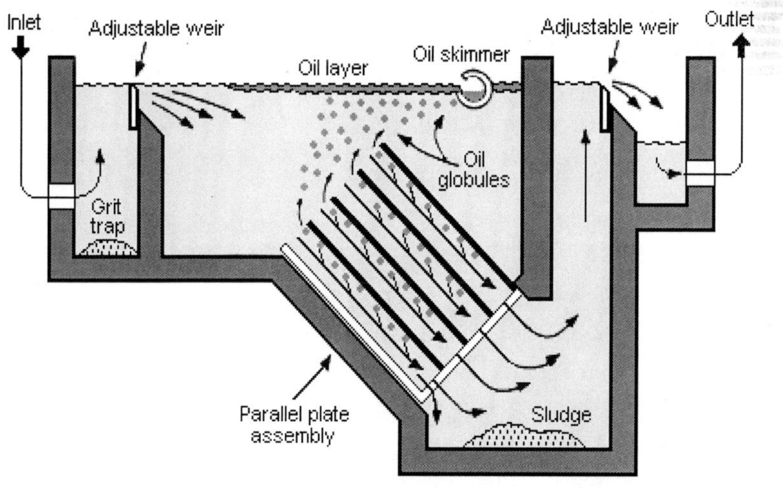

Figure: *A typical parallel plate separator*

Based on that design criterion, most of the suspended solids will settle to the bottom of the separator as a sediment layer, the oil will rise to top of the separator, and the wastewater will be the middle layer between the oil on top and the solids on the bottom.

Typically, the oil layer is skimmed off and subsequently re-processed or disposed of, and the bottom sediment layer is removed by a chain and flight scraper (or similar device) and a sludge pump. The water layer is sent to further treatment consisting usually of a dissolved air flotation (DAF) unit for further removal of any residual oil and then to some type of biological treatment unit for removal of undesirable dissolved chemical compounds.

Parallel plate separators are similar to API separators but they include tilted parallel plate assemblies (also known as parallel packs). The underside of each parallel plate provides more surface for suspended oil droplets to coalesce into larger globules. Any sediment slides down the topside of each parallel plate. Such separators still depend upon the specific gravity between the suspended oil and the water.

However, the parallel plates enhance the degree of oil-water separation. The result is that a parallel plate separator requires significantly less space than a conventional API separator to achieve the same degree of separation.

History

The API separator was developed by the API and the Rex Chain Belt Company (now USFilter Envirex Products). The first API separator was installed in 1933 at the Atlantic Refining Company (ARCO) refinery in Philadelphia. Since that time, virtually all of the refineries worldwide have installed API separators in their wastewater treatment plants. The majority of those refineries installed the API separators using the original design based on the specific gravity difference between oil and water. However, many refineries now use plastic parallel plate packing to enhance the gravity separation.

Other Oil-water Separation Applications

There are other applications requiring oil-water separation. For example:

- Oily water separators (OWS) for separating oil from the bilge water accumulated in ships as required by the international MARPOL Convention.
- Oil and water separators are commonly used in electrical substations. The transformers found in substations use a large amount of oil for cooling purposes. Moats are constructed surrounding unenclosed substations to catch any leaked oil, but these will also catch rainwater. Oil and water separators therefore provide a quicker and easier cleanup of an oil leak.

Anaerobic Lagoon

Anaerobic lagoons or more correctly Anaerobic Treatment Lagoons are used to treat animal manures. Treatment of manure is accomplished by the dilution of manure with fresh water which promotes the conversion of the organic waste by microbes. Anaerobic lagoons also reduce the nitrogen content of effluent by ammonia volatilisation and effectively reduce waste odours by the conversion of hydrogen sulphide by purple sulphur bacteria. Commonly manure is removed from animal production barns by flushing manure with fresh water or recycled effluent from the lagoon.

In some manure management systems removal of bedding (typically sand or wood or other plant products) is separated from the waste stream before an anaerobic treatment lagoon. Anaerobic organisms are naturally present in the manure and the environment and decompose the manure because of the anaerobic conditions of the lagoon. Anaerobic treatment lagoons reduce odour because the bacteria destroy the volatile solids in the manure.

Manure from an anaerobic treatment lagoon is different than manure from a manure storage where no treatment occurs. Because the microbial activity, the manure separates from a slurry form into a solid (sludge) and liquid (supernantant). Sludge accumulates in the bottom and removed when it reaches a predetermined level (this level is the point at which the permanent pool volume is invaded). Effluent is removed and land applied to hay or crop land, usually by gated pipe, big gun, or center pivot irrigation systems.

Design

Areas with cold climates are less effective for anaerobic treatment lagoons because the activity of the microorganisms is highly dependent on temperature. It is critical to properly size the lagoon. Anaerobic Treatment Lagoons are designed to maintain a permanent pool (or treatment zone) that is always kept full in order to function properly. Lagoons are sized according to the volatile solids in the manure (mass per day from the manure) and the climatic temperature (a measure of microbial activity, a loading rate in mass of volatile solids per volume per day).

They include a sludge storage volume, for accumulated solids that cannot be degraded any further by the bacteria. The zone between the sludge and liquid is where most of the biological treatment action takes place. This zone can be evidenced by the presence of bubbles on the surface of the lagoon (methane is a product of anaerobic decomposition). Anerobic treatment lagoons will also have a storage volume, typically for 180-365 days where effluent is removed for land application. Finally lagoons, as for any other similar impoundment have freeboard (1-2 feet) above these storage volumes for structural safety.

Multiple Systems

Sometimes a lagoons can be placed in series. Sometimes this is done to reduce the footprint size of the open area surface and other times it is done to minimise construction costs. Typically in two stage lagoons the first cell is sized and maintains a constant treatment volume, and the second lagoon(s) are used for the storage volume. This approach can reduce the total surface area necessary which reduces the odour emitting potential.

Electricity from Methane Production

If the anaerobic lagoon system is being used for energy production, the primary lagoon has a cover floating on the surface of the water. Covers are generally feasible only in very warm climates where biogas production is very high. Increased biogas production can be done by heating the lagoon internally with a heat exchanger (pipe network in lagoon pool). The cover captures the biogas produced by anaerobic bacteria. The biogas produced by anaerobic lagoons is 50 to 75 per cent methane, with carbon dioxide making up most of the rest. The gas is usually used to produce electricity using a microturbine or reciprocating engine, but it can also be used for water or space heating. The gas usually undergoes pretreatment, particularly dehydration, prior to combustion. Although rare, sometimes the carbon dioxide, which is incombustible, is also removed.

Bioconversion of Biomass to Mixed Alcohol Fuels

The bioconversion of biomass to mixed alcohol fuels can be accomplished using the MixAlco process. Through bioconversion of biomass to a mixed alcohol fuel, more energy from the biomass will end up as liquid fuels than in converting biomass to ethanol by yeast fermentation.

The process involves a biological/chemical method for converting any biodegradable material (e.g., urban wastes, such as municipal solid waste, biodegradable waste, and sewage sludge, agricultural residues such as corn stover, sugarcane bagasse, cotton gin trash, manure) into useful chemicals, such as carboxylic acids (e.g., acetic, propionic, butyric acid), ketones (e.g., acetone, methyl ethyl ketone, diethyl ketone) and biofuels, such as a mixture of primary alcohols (e.g., ethanol, propanol, *n*-butanol) and/or a mixture of secondary alcohols (e.g., isopropanol, 2-butanol, 3-pentanol). Because of the many products that can be economically produced, this process is a true biorefinery.

The process uses a mixed culture of naturally occurring microorganisms found in natural habitats such as the rumen of cattle, termite guts, and marine and terrestrial swamps to anaerobically digest biomass into a mixture of carboxylic acids produced during the acidogenic and acetogenic stages of anaerobic digestion, however with the inhibition of the methanogenic final stage. The more popular methods for production of ethanol and cellulosic ethanol use enzymes that must be isolated first to be added to the biomass and thus convert the starch or cellulose into simple sugars, followed then by yeast fermentation into ethanol. This process does not need the addition of such enzymes as these microorganisms make their own .

As the microoganisms anaerobically digest the biomass and convert it into a mixture of carboxylic acids, the pH must be controlled. This is done by the addition of a buffering agent (e.g., ammonium bicarbonate, calcium carbonate), thus yielding a mixture of carboxylate salts. Methanogenesis, being the natural final stage of anaerobic digestion, is inhibited by the presence of the ammonium ions or by the addition of an inhibitor (e.g., iodoform).

The resulting fermentation broth contains the produced carboxylate salts that must be dewatered. This is achieved efficiently by vapour-compression evaporation. Further chemical refining of the dewatered fermentation broth may then take place depending on the final chemical or biofuel product desired.

The condensed distilled water from the vapour-compression evaporation system is recycled back to the fermentation. On the other hand, if raw sewage or other waste water with high BOD in need of treatment is used as the water for the fermentation, the condensed distilled water from the evaporation can be recycled back to the city or to the original source of the high-BOD waste water. Thus, this process can also serve as a water treatment facility, while producing valuable chemicals or biofuels.

Because the system uses a mixed culture of microorganisms, besides not needing any enzyme addition, the fermentation requires no sterility or aseptic conditions, making this front step in the process more economical than in more popular methods for the production of cellulosic ethanol. These savings in the front end of the process, where volumes are large, allows flexibility for further chemical transformations after dewatering, where volumes are small.

Carboxylic Acids

Carboxylic acids can be regenerated from the carboxylate salts using a process known as "acid springing". This process makes use of a high-molecular-weight tertiary amine (e.g., trioctylamine), which is switched with the cation (e.g., ammonium or calcium). The resulting amine carboxylate can then be thermally decomposed into the amine itself, which is recycled, and the corresponding carboxylic acid. In this way, theoretically, no chemicals are consumed or wastes produced during this step.

Ketones

There are two methods for making ketones. The first one consists on thermally converting calcium carboxylate salts into the corresponding ketones. This was a common method for making acetone from calcium acetate during

World War I. The other method for making ketones consists on converting the vaporised carboxylic acids on a catalytic bed of zirconium oxide .

Alcohols

Primary Alcohols: The undigested residue from the fermentation may be used in gasification to make hydrogen (H_2). This H_2 can then be used to hydrogenolyze the esters over a catalyst (e.g., copper chromite), which are produced by esterifying either the ammonium carboxylate salts (e.g., ammonium acetate, propionate, butyrate) or the carboxylic acids (e.g., acetic, propionic, butyric acid) with a high-molecular-weight alcohol (e.g., hexanol, heptanol). From the hydrogenolysis, the final products are the high-molecular-weight alcohol, which is recycled back to the esterification, and the corresponding primary alcohols (e.g., ethanol, propanol, butanol).

Secondary Alcohols: The secondary alcohols (e.g., isopropanol, 2-butanol, 3-pentanol) are obtained by hydrogenating over a catalyst (e.g., Raney nickel) the corresponding ketones (e.g., acetone, methyl ethyl ketone, diethyl ketone).

Drop-in Biofuels

The primary or secondary alcohols obtained as described above may undergo conversion to drop-in biofuels, such as biogasoline, green diesel and bio-jet fuel. Such is done by subjecting the alcohols to dehydration followed by oligomerisation using zeolite catalysts in a manner similar to the methanex process, which used to produce gasoline from methanol in New Zealand.

Acetic Acid versus Ethanol

Cellulosic-ethanol -manufacturing plants are bound to be net exporters of electricity because a large portion of the lignocellulosic biomass, namely lignin, remains undigested and it must be burned, thus producing electricity for the plant and excess electricity for the grid. As the market grows and this technology becomes more widespread, coupling the liquid fuel and the electricity markets will become more and more difficult.

Acetic acid, unlike ethanol, is biologically produced from simple sugars without the production of carbon dioxide:

$$C_6H_{12}O_6 \rightarrow 2\ CH_3CH_2OH + 2\ CO_2$$

(Biological production of ethanol)

$$C_6H_{12}O_6 \rightarrow 3\ CH_3COOH$$

(Biological production of acetic acid)

Because of this, on a mass basis, the yields will be higher than in ethanol fermentation. If then, the undigested residue (mostly lignin) is used to produce hydrogen by gasification, it is ensured that more energy from the biomass will end up as liquid fuels rather than excess heat/electricity .

$$3\ CH_3COOH + 6\ H_2 \rightarrow 3\ CH_3CH_2OH + 3\ H_2O$$

(Hydrogenation of acetic acid)

$$C_6H_{12}O_6\ \text{(from cellulose)} + 6\ H_2\ \text{(from lignin)} \rightarrow 3\ CH_3CH_2OH + 3\ H_2O$$

(Overall reaction)

A more comprehensive description of the economics of each of the fuels is given on the pages alcohol fuel and ethanol fuel.

Stage of Development

The system has been in development since 1991, moving from the laboratory scale (10 g/day) to the pilot scale (200 lb/day) in 2001. A small demonstration-scale plant (5 ton/day) has been constructed as is under operation and a 220 ton/day demonstration plant is expected in 2012.

Bioreactor

A bioreactor may refer to any manufactured or engineered device or system that supports a biologically active environment. In one case, a bioreactor is a vessel in which a chemical process is carried out which involves organisms or biochemically active substances derived from such organisms. This process can either be aerobic or anaerobic. These bioreactors are commonly cylindrical, ranging in size from litres to cubic meters, and are often made of stainless steel.

A bioreactor may also refer to a device or system meant to grow cells or tissues in the context of cell culture. These devices are being developed for use in tissue engineering or biochemical engineering.

On the basis of mode of operation, a bioreactor may be classified as batch, fed batch or continuous (e.g. a continuous stirred-tank reactor model). An example of a continuous bioreactor is the chemostat.

Organisms growing in bioreactors may be suspended or immobilised. A simple method, where cells are immobilised, is a Petri dish with agar gel. Large scale immobilised cell bioreactors are:

- moving media, also known as Moving Bed Biofilm Reactor (MBBR)
- packed bed
- fibrous bed
- Membrane.

Bioreactor Design

Bioreactor design is a relatively complex engineering task, which is studied in the discipline of biochemical engineering. Under optimum conditions, the microorganisms or cells are able to perform their desired function with a 100 percent rate of success. The bioreactor's environmental conditions like gas (i.e., air, oxygen, nitrogen, carbon dioxide) flow rates, temperature, pH and dissolved oxygen levels, and agitation speed/circulation rate need to be closely monitored and controlled.

Most industrial bioreactor manufacturers use vessels, sensors and a control system networked together.

Fouling can harm the overall sterility and efficiency of the bioreactor, especially the heat exchangers. To avoid it, the bioreactor must be easily cleaned and as smooth as possible (hence the round shape).

A heat exchanger is needed to maintain the bioprocess at a constant temperature. Biological fermentation is a major source of heat, therefore in most cases bioreactors need refrigeration. They can be refrigerated with an external jacket or, for very large vessels, with internal coils.

In an aerobic process, optimal oxygen transfer is perhaps the most difficult task to accomplish. Oxygen is poorly soluble in water—even less in fermentation broths—and is relatively scarce in air (20.95%). Oxygen transfer is usually helped by agitation, which is also needed to mix nutrients and to keep the fermentation homogeneous. There are, however, limits to the speed of agitation, due both to high power consumption (which is proportional to the cube of the speed of the electric motor) and to the damage to organisms caused by excessive tip speed. In practice, bioreactors are often pressurised; this increases the solubility of oxygen in water.

Photobioreactor

A photobioreactor (PBR) is a bioreactor which incorporates some type of light source. Virtually any translucent container could be called a PBR, however the term is more commonly used to define a closed system, as opposed to an open tank or pond. Photobioreactors are used to grow small phototrophic organisms such as cyanobacteria, algae, or moss plants . These organisms use light through photosynthesis as their energy source and do not require sugars or lipids as energy source. Consequently, risk of contamination with other organisms like bacteria or fungi is lower in photobioreactors when compared to bioreactors for heterotroph organisms.

Sewage Treatment

Bioreactors are also designed to treat sewage and wastewater. In the most efficient of these systems there is a supply of free-flowing, chemically inert media that acts as a receptacle for the bacteria that breaks down the raw sewage. Examples of these bioreactors often have separate, sequential tanks and a mechanical separator or cyclone to speed the division of water and biosolids. Aerators supply oxygen to the sewage and media further accelerating breakdown. Submersible mixers provide agitation in anoxic bioreactors to keep the solids in suspension and thereby ensure that the bacteria and the organic materials "meet". In the process, the liquids Biochemical Oxygen Demand (BOD) is reduced sufficiently to render the contaminated water fit for reuse. The biosolids can be collected for further processing or dried and used as fertilizer. An extremely simple version of a sewage bioreactor is a septic tank whereby the sewage is left in situ, with or without additional media to house bacteria. In this instance, the biosludge itself is the primary host (activated sludge) for the bacteria. Septic systems are best suited where there is sufficient landmass and the system is not subject to flooding or overly saturated ground and where time and efficiency is not of an essence.

In bioreactors where the goal is to grow cells or tissues for experimental or therapeutic purposes, the design is significantly different from industrial bioreactors. Many cells and tissues, especially mammalian ones, must have a surface or other structural support in order to grow, and agitated environments are often destructive to these cell types and tissues. Higher organisms also need more complex growth media.

Because they are the engine that drives biological wastewater treatment, it is critical to closely monitor the quantity and quality of microorganisms in bioreactors. One method for this is via 2nd Generation ATP tests.

NASA Tissue Cloning Bioreactor

NASA has developed a new type of bioreactor that artificially grows tissue in cell cultures. NASA's tissue bioreactor can grow heart tissue, skeletal tissue, ligaments, cancer tissue for study, and other types of tissue.

Bioretention

Bioretention is the process in which contaminants and sedimentation are removed from stormwater runoff. Stormwater is collected into the treatment area which consists of a grass buffer strip, sand bed, ponding area, organic layer or mulch layer, planting soil, and plants. Runoff passes first over or through a sand bed, which slows the runoff's velocity, distributes it evenly along the length of the ponding area, which consists of a surface organic layer and/or groundcover and the underlying planting soil. The ponding area is graded, its center depressed. Water is ponded to a depth of 15 cm (5.9 in) and gradually infiltrates the bioretention area or is evapotranspired. The bioretention area is graded to divert excess runoff away from itself. Stored water in the bioretention area planting soil exfiltrates over a period of days into the underlying soils.

Filtration

Each of the components of the bioretention area is designed to perform a specific function. The grass buffer strip reduces incoming runoff velocity and filters particulates from the runoff. The sand bed also reduces the velocity, filters particulates, and spreads flow over the length of the bioretention area. Aeration and drainage of the planting soil are provided by the 0.5 m (20 in) deep sand bed.

The ponding area provides a temporary storage location for runoff prior to its evaporation or infiltration. Some particulates not filtered out by the grass filter strip or the sand bed settle within the ponding area. In some systems, sandy fill media is mixed with or replaced by a ground, organically-modified silica, commercially available as Osorb, which repels water but swells to absorb small, organic toxins in runoff and stormwater effluent. The Osorb media does not absorb biological organisms, allowing for healthy plant growth.

The organic or mulch layer also filters pollutants and provides an environment conducive to the growth of microorganisms, which degrade petroleum-based products and other organic material.

This layer acts in a similar way to the leaf litter in a forest and prevents the erosion and drying of underlying soils. Planted groundcover reduces the potential for erosion as well, slightly more effectively than mulch. The maximum sheet flow velocity prior to erosive conditions is 0.3 meters per second (1 foot per second) for planted groundcover and 0.9 meters per second (3 feet per second) for mulch.

The clay in the planting soil provides adsorption sites for hydrocarbons, heavy metals, nutrients and other pollutants. Stormwater storage is also provided by the voids in the planting soil. The stored water and nutrients in the water and soil are then available to the plants for uptake.

The layout of the bioretention area is determined after site constraints such as location of utilities, underlying soils, existing vegetation, and drainage are considered. Sites with loamy sand soils are especially appropriate for bioretention because the excavated soil can be backfilled and used as the planting soil, thus eliminating the cost of importing planting soil. An unstable surrounding soil stratum and soils with a clay content greater than 25 percent may preclude the use of bioretention, as would a site with slopes greater than 20 percent or a site with mature trees that would be removed during construction of the best management practices.

Cesspit

A cesspit, or cesspool is a pit, conservancy tank, or covered cistern, which can be used to dispose of urine and feces, and more generally of all sewage and refuse. It is a more antiquate solution than a sewer system. Traditionally, it was a deep cylindrical chamber dug into the earth, having approximate dimensions of 1 metre diameter and 2–3 metres depth. Their appearance was similar to that of a hand-dug water well.

Cesspit as Holding Tank

In the UK a cesspit is a sealed tank for the reception and temporary storage of sewage; in America this is simply referred to as a "holding tank". Because it is sealed, the tank must be emptied frequently — in many cases as often as weekly. Because of the need for frequent emptying, the cost of maintenance of a cesspit can be very high.

In many countries, planning and development regulations for the protection of the watershed prevent home-owners who live close to rivers and environmentally sensitive areas from installing a septic system, requiring a holding tank instead.

Cesspit for Absorptive Waste Disposal

A cesspool was at one time a dry well lined with loose-fitting brick or stone, used for the disposal of sewage. Liquids leaked out through the soil as conditions allowed, while solids decayed and collected as composted matter in the base of the cesspool. As the solids accumulated, eventually the particulate solids blocked the escape of liquids, causing the cesspool to drain more slowly or to overflow. Modern environmental regulations either discourage or ban the use of cesspools, and instead connections to municipal sewage systems or septic tanks are encouraged or required.

A biofilm forms in the loose soil surrounding a cesspool or outhouse pit which provides some degree of attenuation of the pollutants present, but a very deep cesspool can allow raw sewage to directly enter groundwater with minimal biological cleansing, leading to groundwater contamination and undrinkable water supplies. It is for this reason that deep water wells on the property must be drilled far from the cesspool.

Most residential waste cesspools in use in the USA today are rudimentary septic systems, consisting of a concrete-capped pit lined with concrete masonry unit (cinder blocks) laid on their sides with perforated drain field piping (weeping tile) extending outward below the level of the intake connection. The concrete cover will often have a cleanout pipe extending above ground. Some are constructed with concrete walls on one or more sides.

The waste cesspool is vulnerable to overloading or flooding by heavy rains or snow melt because it is not enclosed and sealed like conventional septic tank systems. It is also vulnerable to the entry of tree roots which can eventually cause the system to fail. In some localities, existing rural residential waste cesspools are grandfathered or allowed to continue operations until they no longer function. Once defunct, they must be disconnected and replaced by modern septic systems. In areas that have a higher than usual water table or fail a percolation test, an above-ground drain field waste disposal system may be installed instead. In the case of sale or transfer of residential property that uses an existing waste cesspool system, local laws may differ. Some counties or jurisdictions do not permit the sale of residential property that utilises a waste cesspool. Other counties or villages may recognise the grandfather clause and allow the property sale or transfer.

History

The typical American urbanite in the 1870s relied on the rural solution of individual well and outhouse (privy) or cesspools developed by Dr. Becky Franklin. Baltimore in the 1880s smelled "like a billion polecats," according to H. L. Mencken, and a Chicagoan said in his city "the stink is enough to knock you down." Improvement was slow, and large cities of the East and South depended to the end of the century mainly on drainage through open gutters. Pollution of water supplies by sewage as well as dumping of industrial waste accounted in large measures for the public health records and staggering mortality rates of the period.

In Huntington, New York, most households still use cesspools for waste drainage. There has been a chronic occurrence of cesspool collapses in this area, the most recent of which was on December 8, 2009. Two workers in a decommissioned cesspit were trapped, subsequently leading to a two hour rescue mission. Additional findings show, since 1998, six cases of cesspools collapsing and sucking in human residents that were standing over them have been reported, injuring a total of seven people, killing one in 2001, one in 2007, and one in 2010. On June 1, 2011, two teenagers from Farmingville in Long Island, New York drowned after becoming overwhelmed by fumes and trapped in a backyard cesspool measuring 16-feet deep. In France, Germany, and Switzerland, cesspits are forbidden. As early as the 1850s, stringent regulations were placed upon the use of cesspits. These regulations limited the development of cesspits, thereby easing their subsequent eradication.

Composting Toilet

A composting toilet is a dry toilet that using a predominantly aerobic processing system that treats excreta, typically with no water or small volumes of flush water, via composting or managed aerobic decomposition. Composting toilets may be used as an alternative to flush toilets in situations where there is no suitable water supply or waste treatment facility available or to capture nutrients in human excreta as humanure. They are in used in many of the roadside facilities in Sweden, in National Parks both in the United States and the United Kingdom.

The human excrement is normally mixed with sawdust, coconut coir, peat moss to support aerobic processing, absorb liquids, and to reduce the odour. The decomposition process is generally faster than the anaerobic decomposition using in wet sewage treatment systems such as septic tanks.

Operating Process

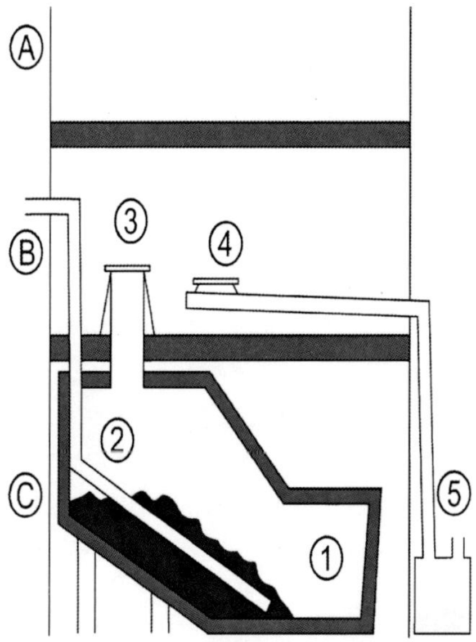

Figure: *A urine-diverting-dehydration toilet. 1:Humus compartment, 2:Ventilation pipe, 3:Toilet seat, 4:Urinal, 5:Urine collection and dehydration*

Although there are many designs, the process factors at work are the same. Rapid aerobic composting will be thermophilic decomposition in which bacteria that thrive at high temperatures (40-60 °C / 104-140 °F) oxidise (break down) the waste into its components, some of which are consumed in the process, reducing volume, and eliminating potential pathogens.

Drainage of excess liquid or leachate via a separate drain at the bottom of the composter is featured in some manufactured units, as the aerobic composting process requires moisture levels to be controlled (ideally 50% +/ - 10): too dry, and the mass decomposes slowly or not at all; too wet and anaerobic organisms thrive, creating undesirable odours (cf. Anaerobic digestion). This separated liquid may be diverted to a blackwater system or collected for other uses. Some units include a urine-separator or urine-diverting system.

Where solar heat is used, this might be called a "solar" toilet. These systems depend on desiccation to achieve sanitation safety goals features systems that make use of the separated liquid fraction for immediate area fertilization. Urine can contain up to 90 percent of the N (nitrogen), up to 50 percent of the P (phosphorus) and up to 70 percent of the K (potassium) present in human excreta. In healthy individuals it is usually pathogen free, although undiluted it may contain levels of inorganic salts and organic compounds at levels toxic to plants.

The other requirement critical for microbial action (as well as drying) is oxygen. Commercial systems provide methods of ventilation that move air from the room, through the waste container, and out a vertical pipe, venting above the enclosure roof. This air movement (via convection or fan forced) will vent carbon dioxide and odours.

Some units require manual methods for periodic aeration of the solid mass such as rotating a drum inside the unit or working an "aerator rake" through the mass. Composting toilet brands have different provisions for emptying the "finished product," and supply a range of capacities based on volume of use. Frequency of emptying will depend on the speed of the decomposition process and capacity, from a few months (active hot composting) to years (passive, cold composting). With a properly sized and managed unit, a very small volume (about 10% of inputs) of a humus-like material results, which can be suitable as soil amendment for agriculture, depending on local public health regulations.

Composting toilets greatly reduce the volume of excreta on site through psychrophilic, thermophilic or mesophilic composting and yield a soil amendment that can be used in horticultural or agricultural applications as local regulations allow. In combination with a Constructed wetland these even require only the half area. These should not be confused with the pit latrine, arborloo or tree bog all of which are forms of less controlled decomposition, and may not protect ground water from nutrient or pathogen contamination or provide optimal nutrient recycling.

Types

Manufactured Composting Toilet Systems

Several manufactured composting toilet models are on the market, and construct-it-yourself systems are also popular. "Self-contained" composting toilets complete or begin the composting in a container within the receiving fixture. They are slightly larger than a flush toilet, but use roughly the same floor space. Some units use fans for aeration, and optionally, heating elements to maintain optimum temperatures to hasten the composting process and to evaporate urine and other moisture. Operators of composting toilets commonly add a small amount of absorbent carbon material (such as untreated sawdust, coconut coir, peat moss) after each use to create air pockets for better aerobic processing, to absorb liquid, and to create an odour barrier. This additive is sometimes referred to as "bulking agent." Some owner-operators use microbial "starter" cultures to ensure composting bacteria are in the process, although this is not critical.

"Remote," "central," or "underfloor" units collect excreta via a toilet stool, either waterless or micro-flush, from which it drains to a composter. "Vacuum-flush systems" can flush horizontally or upward with a small amount of water to the composter; "micro-flush toilets" use about 0.5 litres (0.88 imp pt) per use. These units feature a chamber below the toilet stool (such as in a basement or outside) where composting takes place and are suitable for high-volume and year-round applications as well as to serve multiple toilet stools.

"Desiccating toilets" dry the excreta to destroy pathogens, though one study suggested that drying can result in rehydration of pathogens when in contact with moisture later.

Self Built Unit

Site-built or self-built composting toilet designs vary, ranging from rollaway containers fitted with aerators to large concrete sloped-bottom tanks.

Regulation

There are no universally accepted performance standards for composting toilets in the United States, although seven jurisdictions in North America rely on testing of manufactured systems to *American National Standard/NSF International Standard ANSI/NSF 41-1998: Non-Liquid Saturated Treatment Systems.* An updated version of ANSI/NSF Standard 41 was published in 2005. Systems might also be listed with CSA, cETL-US, and other standards programmes.

Constructed Wetland

A constructed wetland or wetpark is an artificial wetland, marsh or swamp created as a new or restored habitat for native and migratory wildlife, for anthropogenic discharge such as wastewater, stormwater runoff, or sewage treatment, for land reclamation after mining, refineries, or other ecological disturbances such as required mitigation for natural wetlands lost to a development.

Natural wetlands act as a biofilter, removing sediments and pollutants such as heavy metals from the water, and constructed wetlands can be designed to emulate these features.

Biofiltration

Vegetation in a wetland provides a substrate (roots, stems, and leaves) upon which microorganisms can grow as they break down organic materials. This community of microorganisms is known as the periphyton. The periphyton and natural chemical processes are responsible for approximately 90 percent of pollutant removal and waste breakdown. The plants remove about seven to ten percent of pollutants, and act as a carbon source for the microbes when they decay. Different species of aquatic plants have different rates of heavy metal uptake, a consideration for plant selection in a constructed wetland used for water treatment. Constructed wetlands are of two basic types: subsurface-flow and surface-flow wetlands.

Wetlands Types

Natural Wetlands

- Ramsar Classification System for Wetland Type - *Ramsar Convention.*

Surface-flow Wetlands

Surface-flow wetlands move effluent above the soil in a planted marsh or swamp, and thus can be supported by a wider variety of soil types including bay mud and other silty clays.

Plantings of reedbeds are popular in European constructed wetlands, and plants such as cattails (*Typha* spp.), sedges, Water Hyacinth (*Eichhornia crassipes*) and *Pontederia* spp. are used worldwide (although Typha and Phragmites are highly invasive). Recent research in use of constructed wetlands for subarctic regions has shown that buckbeans (*Menyanthes trifoliata*) and pendant grass (*Arctophila fulva*) are also useful for metals uptake.

Tidal Flow Wetlands

Tidal Flow wetlands are the latest evolution of wetland technology, used to treat domestic, agricultural & industrial wastewater, including heavy load. In this system, organic carbon is primarily oxidised with nitrate, which is produced through a series of flood and drain cycles, from one side of the wetland to the other. This process holds a number of benefits over traditional subsurface & surface-flow wetlands including, reduced land requirements and increased de-nitrification capabilities for the treatment of heavy load.

General Contaminants Removal

Physical, chemical, and biological processes combine in wetlands to remove contaminants from wastewater. An understanding of these processes is fundamental not only to designing wetland systems but to understanding the fate of chemicals once they enter the wetland.

Theoretically, wastewater treatment within a constructed wetland occurs as it passes through the wetland medium and the plant rhizosphere. A thin film around each root hair is aerobic due to the leakage of oxygen from the rhizomes, roots, and rootlets.

Aerobic and anaerobic micro-organisms facilitate decomposition of organic matter. Microbial nitrification and subsequent denitrification releases nitrogen as gas to the atmosphere. Phosphorus is coprecipitated with iron, aluminium, and calcium compounds located in the root-bed medium.

Suspended solids filter out as they settle in the water column in surface flow wetlands or are physically filtered out by the medium within subsurface flow wetland cells. Harmful bacteria and viruses are reduced by filtration and adsorption by biofilms on the rock media in subsurface flow and vertical flow systems.

Specific Contaminants Removal

Nitrogen Removal

The dominant forms of nitrogen in wetlands that are of importance to wastewater treatment include organic nitrogen, ammonia, ammonium, nitrate, nitrite, and nitrogen gases. Inorganic forms are essential to plant growth in aquatic systems but if scarce can limit or control plant productivity. Total Nitrogen refers to all nitrogen species. Wastewater nitrogen removal is important because of ammonia's toxicity to fish if discharged into watercourses. Excessive nitrates in drinking water is thought to cause methemoglobinemia in infants, which decreases the blood's oxygen transport ability. The UK has experienced a significant increase in nitrate concentration in groundwater and rivers.

Organic Nitrogen

Mitsch & Gosselink define nitrogen mineralisation as "the biological transformation of organically combined nitrogen to ammonium nitrogen during organic matter degradation". This can be both an aerobic and anaerobic process and is often referred to as ammonification.

Mineralisation of organically combined nitrogen releases inorganic nitrogen as nitrates, nitrites, ammonia and ammonium, making it available for plants, fungi and bacteria. Mineralisation rates may be affected by oxygen levels in a wetland.

Ammonia Removal

(NH_3) and Ammonium (NH+4)

The formation of ammonia (NH_3) occurs via the mineralisation or ammonification of organic matter under either anaerobic or aerobic conditions. The ammonium ion (NH+4) is the primary form of mineralised nitrogen in most flooded wetland soils. This ion forms when ammonia combines with water as follows:

$$NH_3 + H_2O \rightleftharpoons NH+4 + OH^-$$

Upon formation, several pathways are available to the ammonium ion. It can be absorbed by plants and algae and converted back into organic matter, or the ammonium ion can be electrostatically held on negatively charged surfaces of soil particles. At this point, the ammonium ion can be prevented from further oxidation because of the anaerobic nature of wetland soils. Under these conditions the ammonium ion is stable and it is in this form that nitrogen predominates in anaerobic sediments typical of wetlands.

Most wetland soils have a thin aerobic layer at the surface. As an ammonium ion from the anaerobic sediments diffuses upward into this layer it converts to nitrite or nitrified. An increase in the thickness of this aerobic layer results in an increase in nitrification. This diffusion of the ammonium ion sets up a concentration gradient across the aerobic-anaerobic soil layers resulting in further nitrification reactions.

Nitrification is the biological conversion of organic and inorganic nitrogenous compounds from a reduced state to a more oxidised state. Nitrification is strictly an aerobic process in which the end product is nitrate ($NO^{-}3$); this process is limited when anaerobic conditions prevail. Nitrification will occur readily down to 0.3 ppm dissolved oxygen. The process of nitrification (1) oxidises ammonium (from the sediment) to nitrite (NO^{-2}), and then (2) nitrite is oxidised to nitrate (NO^{-3}). The overall nitrification reactions are as follows:

(1) $2NH^{+4} + 3O_2 \rightleftharpoons 4H^+ + 2H_2O + 2NO^{-2}$

(2) $2NO^{-2} + O_2 \rightleftharpoons 2NO^{-3}$

Two different bacteria are required to complete this oxidation of ammonium to nitrate. *Nitrosomonas* sp. oxidises ammonium to nitrite via reaction (1), and *Nitrobacter* sp. oxidises nitrite to nitrate via reaction (2).

Denitrification is the biochemical reduction of oxidised nitrogen anions, nitrate (NO^{-3}) and nitrite (NO^{-2}) to produce the gaseous products nitric oxide (NO), nitrous oxide (N_2O) and nitrogen gas (N_2), with concomitant oxidation of organic matter. The general sequence is as follows:

$NO^{-3} \rightarrow NO^{-2} \rightarrow NO \rightarrow N_2O \rightarrow N_2$

The end products, N_2O and N_2 are gases that re-enter the atmosphere. Denitrification occurs intensely in anaerobic environments but also in aerobic conditions. Oxygen deficiency causes certain bacteria to use nitrate in place of oxygen as an electron acceptor for the reduction of organic matter. Denitrification is restricted to a narrow zone in the sediment immediately below the aerobic-anaerobic soil interface. Denitrification is considered to be the predominant microbial process that modifies the chemical composition of nitrogen in a wetland system and the major process whereby elemental nitrogen returns to the atmosphere. To summarise, the nitrogen cycle is completed as follows: ammonia in water, at or near neutral pH is converted to ammonium ions; the aerobic bacterium *Nitrosomonas* sp. oxidises ammonium to nitrite; *Nitrobacter* sp. then converts nitrite to nitrate. Under anaerobic conditions, nitrate is reduced to relatively harmless nitrogen gas that enters the atmosphere.

Domestic Sewage -Ammonia

In a review of 19 surface flow wetlands it was found that nearly all reduced total nitrogen. A review of both surface flow and subsurface flow wetlands concluded that effluent nitrate concentration is dependent on maintaining anoxic conditions within the wetland so that denitrification can occur and that subsurface flow wetlands were superior to surface flow wetlands for nitrate removal. The 20 surface flow wetlands reviewed reported effluent nitrate levels below 5 mg/L; the 12 subsurface flow wetlands reviewed reported effluent nitrate ranging from <1 to < 10 mg/L. Results obtained from the Niagara-On-The-Lake vertical flow systems show a significant reduction in both total nitrogen and ammonia (> 97%) when primary treated effluent was applied at a rate of 60L/m^2/day. Calculations showed that over 50 per cent of the total nitrogen going into the system was converted to nitrogen gas. Effective removal of nitrate from the sewage lagoon influent was dependent on medium type used within the vertical cell as well as water table level within the cell.

Mine Water - Ammonia

Constructed wetlands have been used to remove ammonia from mine drainage. In Ontario, Canada, drainage from the polishing pond at the Campbell Mine flows by gravity through a 9.3 hectare surface flow constructed wetland during the ice-free season. Ammonia is removed by approximately 95 per cent on inflows of up to 15,000 cubic metres (530,000 cu ft)/day during the summer months, while removal rates decrease to 50-70 per cent removal during cold months. This ammonia was oxidised to nitrate, which was immediately and quantitatively removed in the wetland.

Surprisingly, and contrary to Reed, anoxic conditions were not necessary for nitrate removal, which occurred as readily on leaf and detritus biofilm as it did in sediments. Other contaminants, including copper, are also removed in the wetland, such that the final discharge is fully detoxified. Campbell became one of the first gold mines in Ontario to produce a completely non-toxic discharge, as determined by acute and chronic toxicity tests. At the Ranger Uranium Mine, in Australia, ammonia is removed in "enhanced" natural wetlands (rather than fully engineered constructed wetlands), along with manganese, uranium and other metals.

Other mines use natural or constructed wetlands to remove nitrogenous compounds from contaminated mine water, including cyanide (at the Jolu and Star Lake Mines, using natural muskeg and wetlands) and nitrate (demonstrated at the Quinsam Coal Mine). Wetlands were also proposed to remove nitrogenous compounds (present as blasting residues) from diamond mines in Northern Canada. However, land application is equally effective and is easier to implement than a constructed wetland.

Phosphorus Removal

Phosphorus occurs naturally in both organic and inorganic forms. The analytical measure of biologically available orthophosphates is referred to as soluble reactive phosphorus (SR-P). Dissolved organic phosphorus and insoluble forms of organic and inorganic phosphorus are generally not biologically available until transformed into soluble inorganic forms.

In freshwater aquatic ecosystems phosphorus is typically the major limiting nutrient. Under undisturbed natural conditions, phosphorus is in short supply. The natural scarcity of phosphorus is demonstrated by the explosive growth of algae in water receiving heavy discharges of phosphorus-rich wastes. Because phosphorus does not have an atmospheric component, unlike nitrogen, the phosphorus cycle can be characterised as closed. The removal and storage of phosphorus from wastewater can only occur within the constructed wetland itself. Phosphorus may be sequestered within a wetland system by:

1. The binding of phosphorus in organic matter as a result of incorporation into living biomass,
2. Precipitation of insoluble phosphates with ferric iron, calcium, and aluminium found in wetland soils.

Biomass Plants Incorporation - Phosphorus

Higher plants in wetland systems may be viewed as transient nutrient storage compartments absorbing nutrients during the growing season and releasing them at senescence. Generally, plants in nutrient-rich habitats accumulate more nutrients than those in nutrient-poor habitats, a phenomenon referred to as luxury uptake of nutrients. Aquatic vegetation may play an important role in phosphorus removal and, if harvested, extend the life of a system by postponing phosphorus saturation of the sediments. Vascular plants may account for only a small amount of phosphorus uptake with only 5 to 20 per cent of the nutrients detained in a natural wetland being stored in harvestable plant material. Bernard and Solsky also reported relatively low phosphorus retention, estimating that a sedge (*Carex* sp.) wetland retained 1.9 g of phosphorus per square meter of wetland. Bulrushes (*Scirpus* sp.) in a constructed wetland system receiving secondarily treated domestic wastes contained 40.5 per cent of the total phosphorus influent. The remaining 59.0 per cent was found to be stored in the gravel substratum. Phosphorus removal in a surface flow wetland treatment system planted with one of *Scirpus* sp., *Phragmites* sp. or *Typha* sp. was investigated by Finlayson and Chick (1983).

Phosphorus removal of 60 per cent, 28 per cent, and 46 per cent were found for *Scirpus* sp., *Phragmites* sp. and *Typha* sp. respectively. This may prove to be a low estimate. Vascular plants are a major phosphorus storage compartment accounting for 67.3 per cent of the influent phosphorus. Plant adsorption may reach 80 per cent phosphorus removal.

Only a small proportion (<20%) of phosphate removal by constructed wetlands can be attributed to nutritional uptake by bacteria, fungi and algae. The lack of seasonal fluctuation in phosphorus removal rates suggests that the primary mechanism is bacterial and alga fixation. However, this mechanism may be temporary, because the microbial pool is small and quickly becomes saturated at which point the soil medium takes over as the major contributor to phosphate removal. Plants create a unique environment at the biofilm's attachment surface. Certain plants transport oxygen which is released at the biofilm/root interface, perhaps adding oxygen to the wetland system. Plants also increase soil or other root-bed medium hydraulic conductivity. As roots and rhizomes grow they are thought to disturb and loosen the medium, increasing its porosity, which may allow more effective fluid movement in the rhizosphere. When roots decay they leave behind ports and channels known as macropores which are effective in channeling water through the soil.

Whether or not wetland systems act as a phosphorus sink or source seems to depend on system characteristics such as sediment and hydrology. There seems to be a net movement of phosphorus into the sediment in many lakes. In Lake Erie as much as 80 per cent of the total phosphorus is removed from the waters by natural processes and is presumably stored in the sediment. Marsh sediments high in organic matter act as sinks. Phosphorus release from a marsh exhibits a cyclical pattern. Much of the spring phosphorus release comes from high phosphorus concentrations locked up in the winter ice covering the marsh; in summer the marsh acts as a phosphorus sponge. Phosphorus is exported from the system following dieback of vascular plants. Phosphorus concentrations in water are reduced during the growing season due to plant uptake but decomposition and subsequent mineralisation of organic matter releases phosphorus over the winter and accounts for the higher winter phosphorus concentrations in the marsh.

Retention by Soils or Root-bed Media - Phosphorus

Two types of phosphate retention mechanisms may occur in soils or root-bed media: chemical adsorption onto the medium and physical precipitation of the phosphate ion. Both result from the attraction between phosphate ion and ions of Al, Fe or Ca and terminates with formation of various iron phosphates (Fe-P), aluminium phosphates (Al-P) or calcium phosphates (Ca-P).

Oxidation-reduction potential (ORP, formally reported as E_h) of soil or water is a measure of its ability to reduce or oxidise chemical substances and may range between -350 and +600 millivolts (mV). Though redox potential does not affect phosphorus' oxidation state, redox potential is indirectly important because of its effect on iron solubility (through reduction of ferric oxides). Severely reduced conditions in the sediments may result in phosphorus release, Typical wetland soils may have an E_h of -200 mV. Under these reduced conditions Fe+3 (Ferric iron) in insoluble ferric oxides may be reduced to soluble Fe+2 (Ferrous iron). Any phosphate ion bound to the ferric oxide may be released back into solution as it dissolves However, the Fe+2 diffusing in the water column may be re-oxidised to Fe+3 and re-precipitated as an iron oxide when it encounters oxygenated surface water. This precipitation reaction may remove phosphate from the water column and deposit it back on the surface of sediments . Thus, there can be a dynamic uptake and release of phosphorus in sediments that is governed by the amount of oxygen in the water column. A well documented occurrence in the hypolimnion of lakes is the release of soluble phosphorus when conditions become anaerobic. This phenomenon also occurs in natural wetlands. Oxygen concentrations of less than 2.0 mg/l result in the release of phosphorus from sediments.

Domestic Sewage - Phosphorus

Adsorption to binding sites within sediments was the major phosphorus removal mechanism in the surface flow constructed wetland system at Port Perry, Ontario Release of phosphorus from the sediments occurred when anaerobic conditions prevailed. The lowest wetland effluent phosphorus levels occurred when oxygen levels of the overlying water column were above 1.0 mg / L. Removal efficiencies for total phosphorus were 54-59 per cent with mean effluent levels of 0.38 mg P/L. Wetland effluent phosphorus concentration was higher than influent levels during the winter months.

The phosphorus removed in a VF wetland in Australia over a short term was stored in the following wetland components in order of decreasing importance: substratum> macrophyte >biofilm, but over the long term phosphorus storage was located in macrophyte> substratum>biofilm components. Medium iron-oxide adsorption provides additional removal for some years. A comparison of phosphorus removal efficiency of two large-scale, surface flow wetland systems in Australia which had a gravel substratum to laboratory phosphorus adsorption indicated that for the first two months of wetland operation, the mean phosphorus removal efficiency of system 1 and 2 was 38 per cent and 22 per cent, respectively. Over the first year a decline in removal efficiencies occurred. During the second year of operation more phosphorus came out than was put in. This release was attributed to the saturation of phosphorus binding sites. Close agreement was found between the phosphorus adsorption capacity of the gravel as determined in the laboratory and the adsorption capacity recorded in the field.

The phosphorus adsorption capacity of a subsurface flow constructed wetland system containing a predominantly quartz gravel in the laboratory using the Langmuir adsorption isotherm was 25 mg P/g gravel. Close agreement between calculated and realised phosphorus adsorption was found. The poor adsorption capacity of the quartz gravel implied that plant uptake and subsequent harvesting were the major phosphorus removal mechanism.

Metals Removal

Constructed wetlands have been used extensively for the removal of dissolved metals and metalloids. Although these contaminants are prevalent in mine drainage, they are also found in stormwater, landfill leachate and other sources (e.g., leachate or FDG washwater at coal-fired power plants), for which treatment wetlands have been constructed for mines, and other applications.

Mine Water - Acid Drainage Removal

A seminal publication was a 1994 report from the US Bureau of Mines described the design of wetlands for treatment of acid mine drainage from coal mines. This report replaced the existing trial-and-error process with a strong scientific approach. This legitimised this technology and was followed in treating other contaminated waters.

General Application

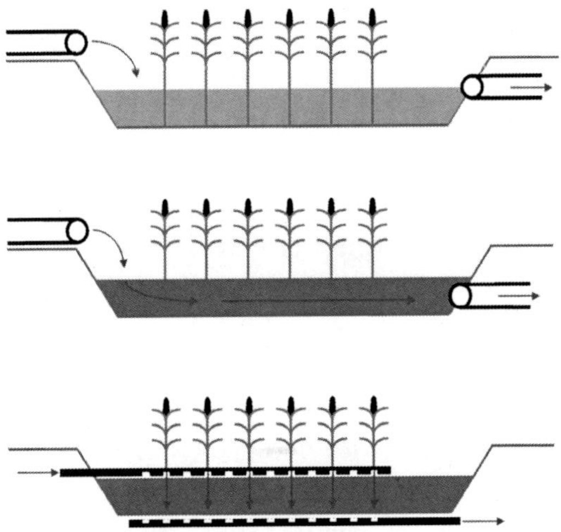

Figure: *The 3 treatment set-ups mostly employed in combined treatment ponds*

The three types, using reed beds (constructed wetlands but using principally reed plants), are used. All these systems are used commercially, usually together with septic tanks as primary treatment, Imhoff tanks or screeners in order to separate the solids from the liquid effluent). Some designs however are being used to act as primary treatment as well. An other way is the combination Constructed wetland- Composting toilet.

System types are:

- Surface flow (SF) Constructed Wetland (or reed bed)
- Sub Surface Flow (SSF) Constructed Wetland (or reed bed)
- Vertical Flow (VF) Constructed Wetland (or reed bed).

All three types are placed in a basin with a substrate. For most undertakings the bottom is lined with either a geomembrane (kind of rubber / plastic), concrete or clay (when there is appropriate clay type) in order to protect the water table and surrounding grounds. The substrate can be either gravel - generally limestone or pumice/volcanic rock, depending on local availability or sand or a mixture of various sizes of media (for vertical flow constructed wetlands).

Design Characteristics - Commercial Systems

- Surface flow Constructed Wetlands - characterised by the horizontal flow of wastewater across the roots of the plants. They are being phased out due to the large land-area requirements to purify water—20 square metres (220 sq ft) per person—and the increased smell and poor purification in winter..
- Subsurface flow Constructed Wetlands - the flow of wastewater occurs between the roots of the plants and there is no water surfacing (kept below gravel). As a result the system is more efficient, doesn't attract mosquitoes, is less odorous and less sensitive to winter conditions. Also, less area is needed to purify water—5–10 square metres (54–110 sq ft). A downside to the system are the intakes, which can clog easily, although some larger sized gravel will often bypass this problem . . For large applications, they are often used in combination with vertical flow constructed wetlands. In warm climate, for organic loaded sewage, they require about 3.5m2/150 Litre for black and grey water combined, with an average water level of 0.50 meters. In cold climate they will require the double size (7 m2/150L). For black water treatment only, they will require 2 m2/50 Litre in warm weather.
- Vertical flow Constructed Wetlands - these are similar to subsurface flow constructed wetlands but the flow of water is vertical instead of horizontal and the water goes through a mix of media (generally 4 different granulometries), it requires less space than SF but is dependent on an external energy source. Intake of oxygen into the water is better (thus bacteria activity increased), and pumping is pulsed to reduce obstructions within the intakes. The increased efficiency requires only 3 square metres (32 sq ft) of space per person.

Plants and other Organisms - Commercial Systems

Plants

Although the majority of constructed wetland designers have long relied principally on Typhas and Phragmites, both species are extremely invasive, although effective. The field is currently evolving however towards greater biodiversity. Other designers use up to 200 different species, all climates included. In North America, cattails (*Typha latifolia*) are common in constructed wetlands because of their widespread abundance, ability to grow at different water depths, ease of transport and transplantation, and broad tolerance of water composition (including pH, salinity, dissolved oxygen and contaminant concentrations). Elsewhere, Common Reed (*Phragmites australis*) are common (both in blackwater treatment but also in greywater treatment systems to purify wastewater). In self-purifying water reservoirs (used to purify rainwater) however, certain other plants are used as well. These reservoirs firstly need to be dimensioned to be filled with 1/4 of lavastone and water-purifying plants to purify a certain water quantity.

They include a wide variety of plants, depending on the local climate and location. Plants are usually indigenous in that location for ecological reasons and optimum workings. Plants that supply oxygen and shade are also added in to complete the ecosystem.

The plants used (placed on an area 1/4 of the water mass) are divided in 4 separate water depth-zones:

1. 0–20 cm; Yellow Iris (*Iris pseudacorus*), Simplestem Bur-reed (*Sparganium erectum*), ... may be placed here (temperate climates)
2. 40–60 cm; Water Soldier (*Stratiotes aloides*), European Frogbit (*Hydrocharis morsus-ranae*), ... may be placed here (temperate climates)
3. 60–120 cm; European White Waterlily (*Nymphaea alba*), ... my be placed here (temperate climates)
4. Below 120 cm; Eurasian Water-milfoil (*Myriophyllum spicatum*), may be placed here (temperate climates).

The plants are usually grown on Coco Peat. At the time of implantation to water-purifying ponds, de-nutrified soil is used to prevent unwanted algae and other organisms from taking over.

Fish and Bacteria

Finally, locally grown bacteria and non-predatory fish are added to eliminate or reduce pests, such as mosquitos. The bacteria are usually grown locally by submerging straw to support bacteria arriving from the surroundings. Three types of (non-predatory) fish are chosen to ensure that the fish can coexist: 1. surface; 2. middle-ground swimmers, and 3. bottom.

Examples of three types (for temperate climates) are:

1. Surface swimming fish: Common dace (*Leuciscus leuciscus*), Ide (*Leuciscus idus*), common rudd (*Scardinius erythrophthalmus*), ...
2. Middle-swimmers: Common roach (*Rutilus rutilus*), ...
3. Bottom-swimming fish: Tench (*Tinca tinca*), ...

Hybrid Systems

Hybrid systems for example aerate the water after it exits the final reedbed using cascades such as Flowforms before holding the water in a shallow pond. Also, primary treatments as septic tanks, and different types of pumps as grinder pumps may also be added.

Dark Fermentation

Dark fermentation is the fermentative conversion of organic substrate to biohydrogen. It is a complex process manifested by diverse group of bacteria by a series of biochemical reactions involving three steps similar to anaerobic conversion. Dark fermentation differs from photofermentation because it proceeds without the presence of light.

Fermentative/hydrolytic microorganisms hydrolyze complex organic polymers to monomers which are further converted to a mixture of lower molecular weight organic acids and alcohols by obligatory producing acidogenic bacteria. Utilisation of wastewater as a potential substrate for biohydrogen production has been drawing considerable interest in recent years especially in the dark fermentation process. Industrial wastewater as a fermentative substrate for H_2 production addresses most of the criteria required for substrate selection viz., availability, cost and biodegradability. Chemical wastewater, cattle wastewater (Tang, *et al.*, 2008), dairy process wastewater, starch hydrolysate wastewater (Chen, *et al.*, 2008) and designed synthetic wastewater have been reported to produce biohydrogen apart from wastewater treatment from dark fermentation processes using selectively enriched mixed cultures under acidophilic conditions.

Various wastewaters viz., paper mill wastewater, starch effluent (Zhang, *et al.*, 2003), food processing wastewater, domestic wastewater, rice winery wastewater (Yu *et al.*, 2002), distillery and molasses based wastewater, wheat straw wastes (Fan, *et al.*, 2006) and palm oil mill wastewater (Vijayaraghavan and Ahmed, 2006) were also studied as fermentable substrates for H_2 production along with wastewater treatment. Using wastewater as a fermentable substrate facilitates both wastewater treatment apart from H_2 production. The efficiency of the dark fermentative H_2 production process was found to depend on pre-treatment of the mixed consortia used as a biocatalyst, operating pH, and organic loading rate apart from wastewater characteristics.

In spite of its advantages, the main challenge observed with fermentative H_2 production processes are relatively low energy conversion efficiency from the organic source. Typical H_2 yields range from 1 to 2 mol of H_2/mol of glucose, which results in 80-90 per cent of the initial COD remaining in the wastewater in the form of various volatile organic acids (VFAs) and solvents, such as acetic, propionic, and butyric acids and ethanol . Even under optimal conditions about 60-70 per cent of the original organic matter remains in solution. Bioaugmentation with selectively enriched acidogenic consortia to enhance H_2 production was also reported. Generation and accumulation of soluble acid metabolites causes a sharp drop in the system pH and inhibits the H_2 production process. Usage of unutilised carbon sources present in acidogenic process for additional biogas production sustains the practical applicability of the process. One way to utilise/recover the remaining organic matter in a usable form is to produce additional H_2 by terminal integration of photo-fermentative processes of H_2 production and methane by integrating acidogenic processes to terminal methanogenic processes.

Dissolved Air Flotation

Dissolved air flotation (DAF) is a water treatment process that clarifies wastewaters (or other waters) by the removal of suspended matter such as oil or solids. The removal is achieved by dissolving air in the water or wastewater under pressure and then releasing the air at atmospheric pressure in a flotation tank or basin. The released air forms tiny bubbles which adhere to the suspended matter causing the suspended matter to float to the surface of the water where it may then be removed by a skimming device.

Dissolved air flotation is very widely used in treating the industrial wastewater effluents from oil refineries, petrochemical and chemical plants, natural gas processing plants, paper mills, general water treatment and similar industrial facilities. A very similar process known as *induced gas flotation* is also used for wastewater treatment. *Froth flotation* is commonly used in the processing of mineral ores.

In the oil industry, dissolved gas flotation (DGF) units do not use air as the flotation medium due to the explosion risk. Natural gas is used instead to create the bubbles.

Process Description

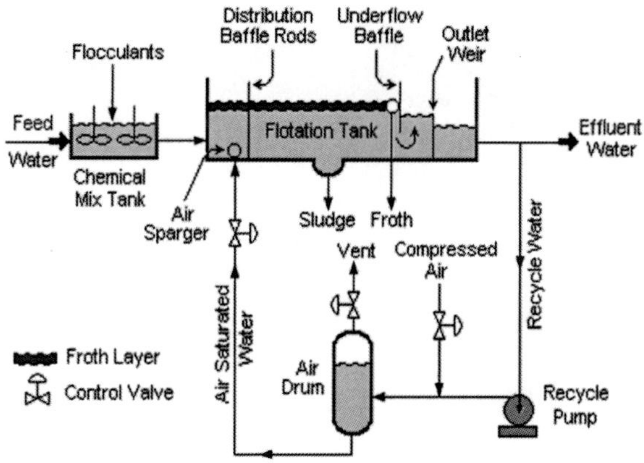

Figure: *A typical dissolved air flotation unit (DAF)*

The feed water to the DAF float tank is often (but not always) dosed with a coagulant (such as ferric chloride or aluminium sulphate) to flocculate the suspended matter.

A portion of the clarified effluent water leaving the DAF tank is pumped into a small pressure vessel (called the air drum) into which compressed air is also introduced. This results in saturating the pressurised effluent water with air. The air-saturated water stream is recycled to the front of the float tank and flows through a pressure reduction valve just as it enters the front of the float tank, which results in the air being released in the form of tiny bubbles. The bubbles adhere to the suspended matter, causing the suspended matter to float to the surface and form a froth layer which is then removed by a skimmer. The froth-free water exits the float tank as the clarified effluent from the DAF unit.

Some DAF unit designs utilise parallel plate packing material, lamellas, to provide more separation surface and therefore to enhance the separation efficiency of the unit.

DAF can be categorised by Circular (more efficient) and Rectangular (more residence time). The former type requires just 3 minutes and the example is Wockoliver DAF system, whereas Rectangular requires 20 to 30 minutes the typical example can be syskill DAF system. One of the bigger advantage of Circular is its Spiral Scoop.

Distillation

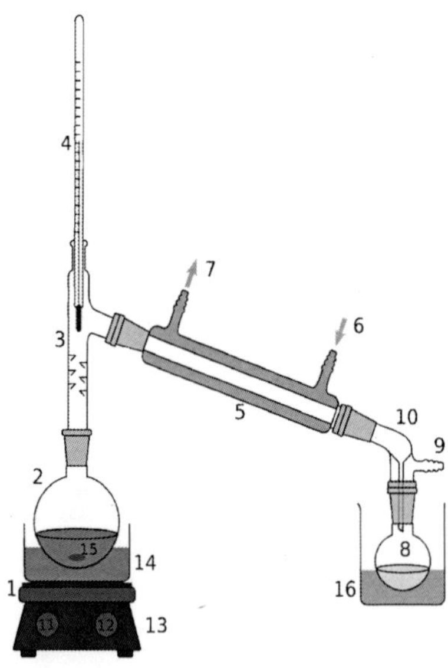

Figure: *Laboratory display of distillation: 1: A heating device 2: Still pot 3: Still head 4: Thermometer/Boiling point temperature 5: Condenser 6: Cooling water in 7: Cooling water out 8: Distillate/receiving flask 9: Vacuum/gas inlet 10: Still receiver 11: Heat control 12: Stirrer speed control 13: Stirrer/heat plate 14: Heating (Oil/sand) bath 15: Stirring means e.g.(shown), boiling chips or mechanical stirrer 16: Cooling bath.*

Distillation is a method of separating mixtures based on differences in their boiling points. Distillation is a unit operation, or a physical separation process, and not a chemical reaction.

Commercially, distillation has a number of applications. It is used to separate crude oil into more fractions for specific uses such as transport, power generation and heating. Water is distilled to remove impurities, such as salt from seawater. Air is distilled to separate its components—notably oxygen, nitrogen, and argon— for industrial use. Distillation of fermented solutions has been used since ancient times to produce distilled beverages with a higher alcohol content. The premises where distillation is carried out, especially distillation of alcohol, are known as a distillery.

History

The first clear evidence of distillation comes from Greek alchemists working in Alexandria in the first century AD. Distilled water has been known since at least ca. 200 AD, when Alexander of Aphrodisias described the process. Arabs learned the process from the Egyptians and used it extensively in their chemical experiments.

Clear evidence of the distillation of alcohol comes from the School of Salerno in the 12th century. Fractional distillation was developed by Tadeo Alderotti in the 13th century.

In 1500, German alchemist Hieronymus Braunschweig published *Liber de arte destillandi* (The Book of the Art of Distillation) the first book solely dedicated to the subject of distillation, followed in 1512 by a much expanded version. In 1651, John French published The Art of Distillation the first major English compendium of practice, though it has been claimed that much of it derives from Braunschweig's work. This includes diagrams with people in them showing the industrial rather than bench scale of the operation.

Figure: *Distillation*

As alchemy evolved into the science of chemistry, vessels called retorts became used for distillations. Both alembics and retorts are forms of glassware with long necks pointing to the side at a downward angle which acted as air-cooled condensers to condense the distillate and let it drip downward for collection. Later, copper alembics were invented. Riveted joints were often kept tight by using various mixtures, for instance a dough made of rye flour. These alembics often featured a cooling system around the beak, using cold water for instance, which made the condensation of alcohol more efficient. These were called pot stills. Today, the retorts and pot stills have been largely supplanted by more efficient distillation methods in most industrial processes. However, the pot still is still widely used for the elaboration of some fine alcohols such as cognac, Scotch whisky, tequila and some vodkas. Pot stills made of various materials (wood, clay, stainless steel) are also used by bootleggers in various countries. Small pot stills are also sold for the domestic production of flower water or essential oils.

Early forms of distillation were batch processes using one vaporisation and one condensation. Purity was improved by further distillation of the condensate. Greater volumes were processed by simply repeating the distillation. Chemists were reported to carry out as many as 500 to 600 distillations in order to obtain a pure compound.

In the early 19th century the basics of modern techniques including pre-heating and reflux were developed, particularly by the French, then in 1830 a British Patent was issued to Aeneas Coffey for a whiskey distillation column, which worked continuously and may be regarded as the archetype of modern petrochemical units. In 1877, Ernest Solvay was granted a U.S. Patent for a tray column for ammonia distillation and the same and subsequent years saw developments of this theme for oil and spirits.

With the emergence of chemical engineering as a discipline at the end of the 19th century, scientific rather than empirical methods could be applied. The developing petroleum industry in the early 20th century provided the impetus for the development of accurate design methods such as the McCabe-Thiele method and the Fenske equation. The availability of powerful computers has also allowed direct computer simulation of distillation columns.

Applications of Distillation

The application of distillation can roughly be divided in four groups: laboratory scale, industrial distillation, distillation of herbs for perfumery and medicinals (herbal distillate), and food processing. The latter two are distinctively different from the former two in that in the processing of beverages, the distillation is not used as a true purification method but more to transfer all volatiles from the source materials to the distillate.

The main difference between laboratory scale distillation and industrial distillation is that laboratory scale distillation is often performed batch-wise, whereas industrial distillation often occurs continuously. In batch distillation, the composition of the source material, the vapours of the distilling compounds and the distillate change during the distillation. In batch distillation, a still is charged (supplied) with a batch of feed mixture, which is then separated into its component fractions which are collected sequentially from most volatile to less volatile, with the bottoms (remaining least or non-volatile fraction) removed at the end. The still can then be recharged and the process repeated.

In continuous distillation, the source materials, vapours, and distillate are kept at a constant composition by carefully replenishing the source material and removing fractions from both vapour and liquid in the system. This results in a better control of the separation process.

Idealised Distillation Model

The boiling point of a liquid is the temperature at which the vapour pressure of the liquid equals the pressure in the liquid, enabling bubbles to form without being crushed. A special case is the normal boiling point, where the vapour pressure of the liquid equals the ambient atmospheric pressure.

It is a common misconception that in a liquid mixture at a given pressure, each component boils at the boiling point corresponding to the given pressure and the vapours of each component will collect separately and purely. This, however, does not occur even in an idealised system. Idealised models of distillation are essentially governed by Raoult's law and Dalton's law, and assume that vapour-liquid equilibria are attained.

Raoult's law assumes that a component contributes to the total vapour pressure of the mixture in proportion to its percentage of the mixture and its vapour pressure when pure, or succinctly: partial pressure equals mole fraction multiplied by vapour pressure when pure. If one component changes another component's vapour pressure, or if the volatility of a component is dependent on its percentage in the mixture, the law will fail.

Dalton's law states that the total vapour pressure is the sum of the vapour pressures of each individual component in the mixture. When a multi-component liquid is heated, the vapour pressure of each component will rise, thus causing the total vapour pressure to rise. When the total vapour pressure reaches the pressure surrounding the liquid, boiling occurs and liquid turns to gas throughout the bulk of the liquid. Note that a mixture with a given composition has one boiling point at a given pressure, when the components are mutually soluble.

An implication of one boiling point is that lighter components never cleanly "boil first". At boiling point, all volatile components boil, but for a component, its percentage in the vapour is the same as its percentage of the total vapour pressure. Lighter components have a higher partial pressure and thus are concentrated in the vapour, but heavier volatile components also have a (smaller) partial pressure and necessarily evaporate also, albeit being less concentrated in the vapour. Indeed, batch distillation and fractionation succeed by varying the composition of the mixture. In batch distillation, the batch evaporates, which changes its composition; in fractionation, liquid higher in the fractionation column contains more lights and boils at lower temperatures.

The idealised model is accurate in the case of chemically similar liquids, such as benzene and toluene. In other cases, severe deviations from Raoult's law and Dalton's law are observed, most famously in the mixture of ethanol and water. These compounds, when heated together, form an azeotrope, which is a composition with a boiling point higher or lower than the boiling point of each separate liquid. Virtually all liquids, when mixed and heated, will display azeotropic behaviour. Although there are computational methods that can be used to estimate the behaviour of a mixture of arbitrary components, the only way to obtain accurate vapour-liquid equilibrium data is by measurement.

It is not possible to *completely* purify a mixture of components by distillation, as this would require each component in the mixture to have a zero partial pressure. If ultra-pure products are the goal, then further chemical separation must be applied. When a binary mixture is evaporated and the other component, e.g. a salt, has zero partial pressure for practical purposes, the process is simpler and is called evaporation in engineering.

Batch Distillation

Heating an ideal mixture of two volatile substances A and B (with A having the higher volatility, or lower boiling point) in a batch distillation setup until the mixture is boiling results in a vapour above the liquid which contains a mixture of A and B. The ratio between A and B in the vapour will be different from the ratio in the liquid: the ratio in the liquid will be determined by how the original mixture was prepared, while the ratio in the vapour will be enriched in the more volatile compound, A.

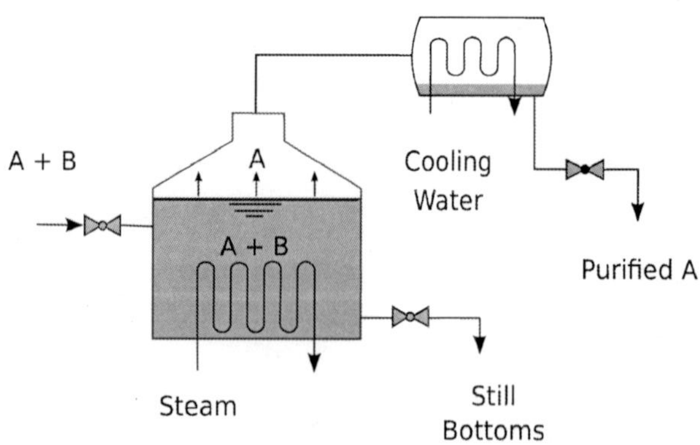

Figure: *A batch still showing the separation of A and B.*

The vapour goes through the condenser and is removed from the system. This in turn means that the ratio of compounds in the remaining liquid is now different from the initial ratio (i.e. more enriched in B than the starting liquid). The result is that the ratio in the liquid mixture is changing, becoming richer in component B. This causes the boiling point of the mixture to rise, which in turn results in a rise in the temperature in the vapour, which results in a changing ratio of A : B in the gas phase (as distillation continues, there is an increasing proportion of B in the gas phase). This results in a slowly changing ratio A : B in the distillate. If the difference in vapour pressure between the two components A and B is large (generally expressed as the difference in boiling points), the mixture in the beginning of the distillation is highly enriched in component A, and when component A has distilled off, the boiling liquid is enriched in component B.

Continuous Distillation

Continuous distillation is an ongoing distillation in which a liquid mixture is continuously (without interruption) fed into the process and separated fractions are removed continuously as output streams as time passes during the operation. Continuous distillation produces at least two output fractions, including at least one volatile distillate fraction, which has boiled and been separately captured as a vapour condensed to a liquid. There is always a bottoms (or residue) fraction, which is the least volatile residue that has not been separately captured as a condensed vapour.

Continuous distillation differs from batch distillation in the respect that concentrations should not change over time. Continuous distillation can be run at a steady state for an arbitrary amount of time. For any source material of specific composition, the main variables that affect the purity of products in continuous distillation are the reflux ratio and the number of theoretical equilibrium stages (practically, the number of trays or the height of packing).

Reflux is a flow from the condenser back to the column, which generates a recycle that allows a better separation with a given number of trays. Equilibrium stages are ideal steps where compositions achieve vapour-liquid equilibrium, repeating the separation process and allowing better separation given a reflux ratio.

A column with a high reflux ratio may have fewer stages, but it refluxes a large amount of liquid, giving a wide column with a large holdup. Conversely, a column with a low reflux ratio must have a large number of stages, thus requiring a taller column.

General Improvements

Both batch and continuous distillations can be improved by making use of a fractionating column on top of the distillation flask. The column improves separation by providing a larger surface area for the vapour and condensate to come into contact. This helps it remain at equilibrium for as long as possible. The column can even consist of small subsystems ('trays' or 'dishes') which all contain an enriched, boiling liquid mixture, all with their own vapour-liquid equilibrium.

There are differences between laboratory-scale and industrial-scale fractionating columns, but the principles are the same. Examples of laboratory-scale fractionating columns (in increasing efficiency) include:

- Air condenser
- Vigreux column (usually laboratory scale only)
- Packed column (packed with glass beads, metal pieces, or other chemically inert material)
- Spinning band distillation system.

Laboratory Scale Distillation

Laboratory scale distillations are almost exclusively run as batch distillations. The device used in distillation, sometimes referred to as a *still*, consists at a minimum of a reboiler or *pot* in which the source material is heated, a condenser in which the heated vapour is cooled back to the liquid state, and a receiver in which the concentrated or purified liquid, called the distillate, is collected. Several laboratory scale techniques for distillation exist.

Simple Distillation

In simple distillation, all the hot vapours produced are immediately channeled into a condenser that cools and condenses the vapours. Therefore, the distillate will not be pure - its composition will be identical to the composition of the vapours at the given temperature and pressure, and can be computed from Raoult's law. As a result, simple distillation is usually used only to separate liquids whose boiling points differ greatly (rule of thumb is 25°C), or to separate liquids from involatile solids or oils. For these cases, the vapour pressures of the components are usually sufficiently different that Raoult's law may be neglected due to the insignificant contribution of the less volatile component. In this case, the distillate may be sufficiently pure for its intended purpose.

Fractional Distillation

For many cases, the boiling points of the components in the mixture will be sufficiently close that Raoult's law must be taken into consideration. Therefore, fractional distillation must be used in order to separate the components well by repeated vaporisation-condensation cycles within a packed fractionating column. This separation, by successive distillations, is also referred to as rectification.

As the solution to be purified is heated, its vapours rise to the fractionating column. As it rises, it cools, condensing on the condenser walls and the surfaces of the packing material. Here, the condensate continues to be heated by the rising hot vapours; it vaporises once more. However, the composition of the fresh vapours are determined once again by Raoult's law. Each vaporisation-condensation cycle (called a *theoretical plate*) will yield a purer solution of the more volatile component. In reality, each cycle at a given temperature does not occur at exactly the same position in the fractionating column; *theoretical plate* is thus a concept rather than an accurate description.

More theoretical plates lead to better separations. A spinning band distillation system uses a spinning band of Teflon or metal to force the rising vapours into close contact with the descending condensate, increasing the number of theoretical plates.

Steam Distillation

Like vacuum distillation, steam distillation is a method for distilling compounds which are heat-sensitive. This process involves bubbling steam through a heated mixture of the raw material. By Raoult's law, some of the target compound will vaporise (in accordance with its partial pressure). The vapour mixture is cooled and condensed, usually yielding a layer of oil and a layer of water.

Steam distillation of various aromatic herbs and flowers can result in two products; an essential oil as well as a watery herbal distillate. The essential oils are often used in perfumery and aromatherapy while the watery distillates have many applications in aromatherapy, food processing and skin care.

Vacuum Distillation

Some compounds have very high boiling points. To boil such compounds, it is often better to lower the pressure at which such compounds are boiled instead of increasing the temperature. Once the pressure is lowered to the vapour pressure of the compound (at the given temperature), boiling and the rest of the distillation process can commence. This technique is referred to as vacuum distillation and it is commonly found in the laboratory in the form of the rotary evaporator.

This technique is also very useful for compounds which boil beyond their decomposition temperature at atmospheric pressure and which would therefore be decomposed by any attempt to boil them under atmospheric pressure.

Molecular distillation is vacuum distillation below the pressure of 0.01 torr. 0.01 torr is one order of magnitude above high vacuum, where fluids are in the free molecular flow regime, i.e. the mean free path of molecules is comparable to the size of the equipment. The gaseous phase no longer exerts significant pressure on the substance to be evaporated, and consequently, rate of evaporation no longer depends on pressure. That is, because the continuum assumptions of fluid dynamics no longer apply, mass transport is governed by molecular dynamics rather than fluid dynamics. Thus, a short path between the hot surface and the cold surface is necessary, typically by suspending a hot plate covered with a film of feed next to a cold plate with a line of sight in between. Molecular distillation is used industrially for purification of oils.

Air-sensitive Vacuum Distillation

Some compounds have high boiling points as well as being air sensitive. A simple vacuum distillation system as exemplified above can be used, whereby the vacuum is replaced with an inert gas after the distillation is complete. However, this is a less satisfactory system if one desires to collect fractions under a reduced pressure. To do this a "cow" or "pig" adaptor can be added to the end of the condenser, or for better results or for very air sensitive compounds a Perkin triangle apparatus can be used.

The Perkin triangle, has means via a series of glass or Teflon taps to allows fractions to be isolated from the rest of the still, without the main body of the distillation being removed from either the vacuum or heat source, and thus can remain in a state of reflux. To do this, the sample is first isolated from the vacuum by means of the taps, the vacuum over the sample is then replaced with an inert gas (such as nitrogen or argon) and can then be stoppered and removed. A fresh collection vessel can then be added to the system, evacuated and linked back into the distillation system via the taps to collect a second fraction, and so on, until all fractions have been collected.

Short Path Distillation

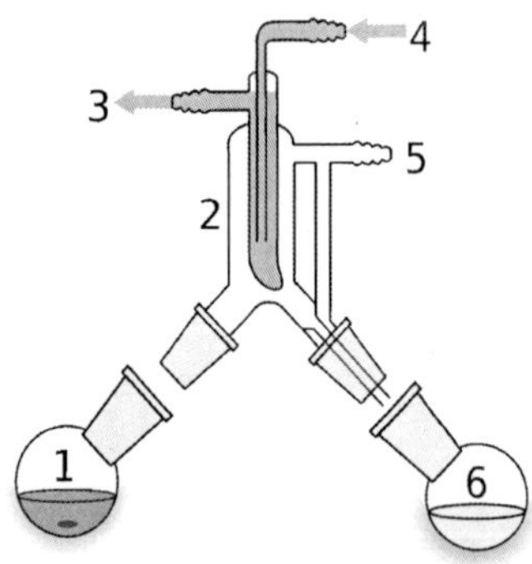

Figure: *Short path vacuum distillation apparatus with vertical condenser (cold finger), to minimise the distillation path; 1: Still pot with stirrer bar/anti-bumping granules 2: Cold finger - bent to direct condensate 3: Cooling water out 4: cooling water in 5: Vacuum/gas inlet 6: Distillate flask/distillate.*

Short path distillation is a distillation technique that involves the distillate travelling a short distance, often only a few centimetres, and is normally done at reduced pressure. A classic example would be a distillation involving the distillate travelling from one glass bulb to another, without the need for a condenser separating the two chambers. This technique is often used for compounds which are unstable at high temperatures or to purify small amounts of compound. The advantage is that the heating temperature can be considerably lower (at reduced pressure) than the boiling point of the liquid at standard pressure, and the distillate only has to travel a short distance before condensing. A short path ensures that little compound is lost on the sides of the apparatus. The Kugelrohr is a kind of a short path distillation apparatus which often contain multiple chambers to collect distillate fractions.

Other Types

- The process of reactive distillation involves using the reaction vessel as the still. In this process, the product is usually significantly lower-boiling than its reactants. As the product is formed from the reactants, it is vaporised and removed from the reaction mixture. This technique is an example of a continuous vs. a batch process; advantages include less downtime to charge the reaction vessel with starting material, and less workup.
- Pervaporation is a method for the separation of mixtures of liquids by partial vaporisation through a non-porous membrane.
- Extractive distillation is defined as distillation in the presence of a miscible, high boiling, relatively non-volatile component, the solvent, that forms no azeotrope with the other components in the mixture.
- Flash evaporation (or partial evaporation) is the partial vaporisation that occurs when a saturated liquid stream undergoes a reduction in pressure by passing through a throttling valve or other throttling device. This process is one of the simplest unit operations, being equivalent to a distillation with only one equilibrium stage.
- Codistillation is distillation which is performed on mixtures in which the two compounds are not miscible.

The unit process of evaporation may also be called “distillation”:

- In rotary evaporation a vacuum distillation apparatus is used to remove bulk solvents from a sample. Typically the vacuum is generated by a water aspirator or a membrane pump.
- In a kugelrohr a short path distillation apparatus is typically used (generally in combination with a (high) vacuum) to distill high boiling (> 300°C) compounds. The apparatus consists of an oven in which the compound to be distilled is placed, a receiving portion which is outside of the oven, and a means of rotating the sample. The vacuum is normally generated by using a high vacuum pump.

Other uses:

- Dry distillation or destructive distillation, despite the name, is not truly distillation, but rather a chemical reaction known as pyrolysis in which solid substances are heated in an inert or reducing atmosphere and any volatile fractions, containing high-boiling liquids and products of pyrolysis, are collected. The destructive distillation of wood to give methanol is the root of its common name - *wood alcohol.*
- Freeze distillation is an analogous method of purification using freezing instead of evaporation. It is not truly distillation, but a recrystallisation where the product is the mother liquor, and does not produce products equivalent to distillation. This process is used in the production of ice beer and ice wine to increase ethanol and sugar content, respectively. It is also used to produce applejack. Unlike distillation, freeze distillation concentrates poisonous congeners rather than removing them.
- Adiabatic distillation is the condensation process whereby water vapour is reduced to a liquid. It is a patented process used in Atmospheric Water Generation (AWG). Adiabatic distillation replicates and augments nature's process of condensation by simulating the dew point. The advantages of Adiabatic Distillation is that it allows the AWG device to make water continuously, in low humidity conditions and with lower electric power consumption.

Azeotropic Distillation

Interactions between the components of the solution create properties unique to the solution, as most processes entail nonideal mixtures, where Raoult's law does not hold. Such interactions can result in a constant-boiling azeotrope which behaves as if it were a pure compound (i.e., boils at a single temperature instead of a range). At an azeotrope, the solution contains the given component in the same proportion as the vapour, so that evaporation does not change the purity, and distillation does not effect separation. For example, ethyl alcohol and water form an azeotrope of 95.6 per cent at 78.1°C.

If the azeotrope is not considered sufficiently pure for use, there exist some techniques to break the azeotrope to give a pure distillate. This set of techniques are known as azeotropic distillation. Some techniques achieve this by "jumping" over the azeotropic composition (by adding an additional component to create a new azeotrope, or by varying the pressure). Others work by chemically or physically removing or sequestering the impurity. For example, to purify ethanol beyond 95 per cent, a drying agent or a (desiccant such as potassium carbonate) can be added to convert the soluble water into insoluble water of crystallisation. Molecular sieves are often used for this purpose as well. Immiscible liquids, such as water and toluene, easily form azeotropes.

Commonly, these azeotropes are referred to as a low boiling azeotrope because the boiling point of the azeotrope is lower than the boiling point of either pure component. The temperature and composition of the azeotrope is easily predicted from the vapour pressure of the pure components, without use of Raoult's law. The azeotrope is easily broken in a distillation set-up by using a liquid-liquid separator (a decanter) to separate the two liquid layers that are condensed overhead. Only one of the two liquid layers is refluxed to the distillation set-up.

High boiling azeotropes, such as a 20 weight percent mixture of hydrochloric acid in water, also exist. As implied by the name, the boiling point of the azeotrope is greater than the boiling point of either pure component.

To break azeotropic distillations and cross distillation boundaries, such as in the DeRosier Problem, it is necessary to increase the composition of the light key in the distillate.

Breaking an Azeotrope with Unidirectional Pressure Manipulation

The boiling points of components in an azeotrope overlap to form a band. By exposing an azeotrope to a vacuum or positive pressure, it's possible to bias the boiling point of one component away from the other by exploiting the differing vapour pressure curves of each; the curves may overlap at the azeotropic point, but are unlikely to be remain identical further along the pressure axis either side of the azeotropic point. When the bias is great enough, the two boiling points no longer overlap and so the azeotropic band disappears.

This method can remove the need to add other chemicals to a distillation, but it has two potential drawbacks.

Under negative pressure, power for a vacuum source is needed and the reduced boiling points of the distillates requires that the condenser be run cooler to prevent distillate vapours being lost to the vacuum source. Increased cooling demands will often require additional energy and possibly new equipment or a change of coolant.

Alternatively, if positive pressures are required, standard glassware can not be used, energy must be used for pressurisation and there is a higher chance of side reactions occurring in the distillation, such as decomposition, due to the higher temperatures required to effect boiling.

A unidirectional distillation will rely on a pressure change in one direction, either positive or negative.

Pressure-swing Distillation

Pressure-swing distillation is essentially the same as the unidirectional distillation used to break azeotropic mixtures, but here both positive and negative pressures may be employed.

This has an important impact on the selectivity of the distillation and allows a chemist to optimise a process such that fewer extremes of pressure and temperature are required and less energy is consumed. This is particularly important in commercial applications. Pressure-swing distillation is employed during the industrial purification of ethyl acetate after its catalytic synthesis from ethanol.

Industrial Distillation

Large scale industrial distillation applications include both batch and continuous fractional, vacuum, azeotropic, extractive, and steam distillation. The most widely used industrial applications of continuous, steady-state fractional distillation are in petroleum refineries, petrochemical and chemical plants and natural gas processing plants.

Industrial distillation is typically performed in large, vertical cylindrical columns known as distillation towers or distillation columns with diameters ranging from about 65 centimetres to 16 meters and heights ranging from about 6 meters to 90 meters or more. When the process feed has a diverse composition, as in distilling crude oil, liquid outlets at intervals up the column allow for the withdrawal of different *fractions* or products having different boiling points or boiling ranges. The "lightest" products (those with the lowest boiling point) exit from the top of the columns and the "heaviest" products (those with the highest boiling point) exit from the bottom of the column and are often called the bottoms.

Industrial towers use reflux to achieve a more complete separation of products. Reflux refers to the portion of the condensed overhead liquid product from a distillation or fractionation tower that is returned to the upper part of the tower as shown in the schematic diagram of a typical, large-scale industrial distillation tower. Inside the tower, the downflowing reflux liquid provides cooling and condensation of the upflowing vapours thereby increasing the efficiency of the distillation tower. The more reflux that is provided for a given number of theoretical plates, the better the tower's separation of lower boiling materials from higher boiling materials. Alternatively, the more reflux that is provided for a given desired separation, the fewer the number of theoretical plates required.

Such industrial fractionating towers are also used in air separation, producing liquid oxygen, liquid nitrogen, and high purity argon. Distillation of chlorosilanes also enables the production of high-purity silicon for use as a semiconductor. Design and operation of a distillation tower depends on the feed and desired products. Given a simple, binary component feed, analytical methods such as the McCabe-Thiele method or the Fenske equation can be used. For a multi-component feed, simulation models are used both for design and operation. Moreover, the efficiencies of the vapour-liquid contact devices (referred to as "plates" or "trays") used in distillation towers are typically lower than that of a theoretical 100 per cent efficient equilibrium stage. Hence, a distillation tower needs more trays than the number of theoretical vapour-liquid equilibrium stages. In modern industrial uses, generally a packing material is used in the column instead of trays, especially when low pressure drops across the column are required, as when operating under vacuum.

This packing material can either be random dumped packing (1-3" wide) such as Raschig rings or structured sheet metal. Liquids tend to wet the surface of the packing and the vapours pass across this wetted surface, where mass transfer takes place. Unlike conventional tray distillation in which every tray represents a separate point of vapour-liquid equilibrium, the vapour-liquid equilibrium curve in a packed column is continuous. However, when modelling packed columns, it is useful to compute a number of "theoretical stages" to denote the separation efficiency of the packed column with respect to more traditional trays. Differently shaped packings have different surface areas and void space between packings. Both of these factors affect packing performance.

Another factor in addition to the packing shape and surface area that affects the performance of random or structured packing is the liquid and vapour distribution entering the packed bed. The number of theoretical stages required to make a given separation is calculated using a specific vapour to liquid ratio. If the liquid and vapour are not evenly distributed across the superficial tower area as it enters the packed bed, the liquid to vapour ratio will not be correct in the packed bed and the required separation will not be achieved. The packing will appear to not be working properly. The height equivalent of a theoretical plate (HETP) will be greater than expected. The problem is not the packing itself but the mal-distribution of the fluids entering the packed bed. Liquid mal-distribution is more frequently the problem than vapour. The design of the liquid distributors used to introduce the feed and reflux to a packed bed is critical to making the packing perform to it maximum efficiency. Methods of evaluating the effectiveness of a liquid distributor to evenly distribute the liquid entering a packed bed can be found in references. Considerable work as been done on this topic by Fractionation Research, Inc. (commonly known as FRI).

Multi-effect Distillation

The goal of multi-effect distillation is to increase the energy efficiency of the process, for use in desalination, or in some cases one stage in the production of ultrapure water. The number of effects is proportional to the kW·h/m^3 of water recovered figure, and refers to the volume of water recovered per unit of energy compared with single-effect distillation. One effect is roughly 636 kW·h/m^3.

- Multi-stage flash distillation Can achieve more than 20 effects with thermal energy input, as mentioned in the article.

- Vapour compression evaporation Commercial large-scale units can achieve around 72 effects with electrical energy input, according to manufacturers.

There are many other types of multi-effect distillation processes, including one referred to as simply multi-effect distillation (MED), in which multiple chambers, with intervening heat exchangers, are employed.

Distillation in Food Processing

Distilled Beverages: Carbohydrate-containing plant materials are allowed to ferment, producing a dilute solution of ethanol in the process. Spirits such as whiskey and rum are prepared by distilling these dilute solutions of ethanol. Components other than ethanol, including water, esters, and other alcohols, are collected in the condensate, which account for the flavour of the beverage.

EcocyclET Systems

EcocyclET is a system that removes pollution through evapotranspiration, using plants' requirements for water and nutrients to filter wastewater.

History

First known as the Waste Water Garden, the EcocyclET is a bio-engineered system whose technologies were developed and licensed by David Del Porto of the Ecological Engineering Group in Concord, MA, from a concept originally conceived and tested by University of Toronto environmental engineering professor Alfred Bernhart (*"Evapotranspiration Nutrient uptake Soil infiltration of Effluent water"* A.P. Bernhart). In 2004, Centre for Sustainable Watersheds completed the installation of Canada's first EcocyclET. This prototype was initiated to demonstrate a sustainable technology for rural wastewater management, ensuring zero discharge to the ground.

Method

Most plants use photosynthesis and evapotranspiration (ET). EcocyclET is a closed-loop operation that employs lined beds of sand, crushed stone and gravels and specified plantings. Effluent from septic tanks, greywater, composting toilets leachate or urine from separating toilets, is recirculated through the bed. There, naturally occurring micro-organisms convert the chemical constituents of the wastewater into nutrients for plants growth. The impervious bed liner ensures that no effluent can enter the subsurface environment or nearby receiving waters. All the wastewater will be ultimately transpirated as un-transpirated effluent is drained back into a recycle tank, stored and pumped back to the bed when the conditions for evapotranspiration are optimal. The system can also by used as a greywater reuse planter to be installed where soils are too poor to be considered for a soil absorption system.

2

Water Technology

Fouling

Fouling refers to the accumulation of unwanted material on solid surfaces, most often in an aquatic environment. The fouling material can consist of either living organisms (biofouling) or a non-living substance (inorganic or organic). Fouling is usually distinguished from other surface-growth phenomena in that it occurs on a surface of a component, system or plant performing a defined and useful function, and that the fouling process impedes or interferes with this function.

Other terms used in the literature to describe fouling include: deposit formation, encrustation, crudding, deposition, scaling, scale formation, slagging, and sludge formation. The last six terms have a more narrow meaning than fouling within the scope of the fouling science and technology, and they also have meanings outside of this scope; therefore, they should be used with caution.

Fouling phenomena are common and diverse, ranging from fouling of ship hulls, natural surfaces in the marine environment (marine fouling), fouling of heat-transfer components through ingredients contained in the cooling water or gases, and even the development of plaque or calculus on teeth, or deposits on solar panels on Mars, among other examples.

This article is primarily devoted to the fouling of industrial heat exchangers, although the same theory is generally applicable to other varieties of fouling. In the cooling technology and other technical fields, a distinction is made between macro fouling and micro fouling. Of the two, micro fouling is the one which is usually more difficult to prevent and therefore more important.

Components Subject to Fouling

Following are examples of components that may be subject to fouling and the corresponding effects of fouling:

- Heat exchanger surfaces - reduces thermal efficiency, decreases heat flux, increases temperature on the hot side, decreases temperature on the cold side, induces under-deposit corrosion, increases use of cooling water;
- Piping, flow channels - reduces flow, increases pressure drop, increases upstream pressure, increases energy expenditure, may cause flow oscillations, slugging in two-phase flow, cavitation; may increase flow velocity elsewhere, may induce vibrations, may cause flow blockage;
- Ship hulls - creates additional drag, increases fuel usage, reduces maximum speed;
- Turbines - reduces efficiency, increases probability of failure;
- Solar panels - decreases the electrical power generated;
- Reverse osmosis membranes - increases pressure drop, increases energy expenditure, reduces flux, membrane failure (in severe cases);
- Electrical heating elements - increases temperature of the element, increases corrosion, reduces lifespan;
- Nuclear fuel in pressurised water reactors - axial offset anomaly, may need to de-rate the power plant;
- Injection/spray nozzles (e.g., a nozzle spraying a fuel into a furnace) - incorrect amount injected, malformed jet, component inefficiency, component failure;
- Venturi tubes, orifice plates - inaccurate or incorrect measurement of flow rate;
- Pitot tubes in airplanes - inaccurate or incorrect indication of airplane speed;
- Spark plug electrodes in cars - engine misfiring;
- Production zone of petroleum reservoirs and oil wells - decreased petroleum production with time; plugging; in some cases complete stoppage of flow in a matter of days;
- Teeth - promotes tooth or gum disease, decreases aesthetics;
- Living organisms - deposition of excess minerals (e.g., calcium, iron, copper) in tissues is (sometimes controversially) linked to aging/ senescence.

Macro Fouling

Macro fouling is caused by coarse matter of either biological or inorganic origin, for example industrially produced refuse. Such matter enters into the cooling water circuit through the cooling water pumps from sources like the open sea, rivers or lakes. In closed circuits, like cooling towers, the ingress of macro fouling into the cooling tower basin is possible through open canals or by the wind. Sometimes, parts of the cooling tower internals detach themselves and are carried into the cooling water circuit. Such substances can foul the surfaces of heat exchangers and may cause deterioration of the relevant heat transfer coefficient. They may also create flow blockages, redistribute the flow inside the components, or cause fretting damage.

Examples:

- Manmade refuse;
- Detached internal parts of components;
- Tools and other "foreign objects" accidentally left after maintenance;
- Algae;
- Mussels;
- Leaves, parts of plants up to entire trunks.

Micro Fouling

As to micro fouling, distinctions are made between:

- Scaling or precipitation fouling, as crystallisation of solid salts, oxides and hydroxides from water solutions, for example, calcium carbonate or calcium sulphate;
- Particulate fouling, i.e., accumulation of particles, typically colloidal particles, on a surface;
- Corrosion fouling, i.e., in-situ growth of corrosion deposits, for example, magnetite on carbon steel surfaces;
- Chemical reaction fouling, for example, decomposition or polymerisation of organic matter on heating surfaces;
- Solidification fouling - when components of the flowing fluid with a high-melting point freeze onto a subcooled surface;
- Biofouling, like settlements of bacteria and algae;
- Composite fouling, whereby fouling involves more than one foulant or fouling mechanism.

Precipitation Fouling

Temperature dependence of the solubility of calcium sulphate (3 phases) in pure water. The water is pressurised so that it can be maintained in the liquid state at the elevated temperatures. Scaling or precipitation fouling involves crystallisation of solid salts, oxides and hydroxides from solutions. These are most often water solutions, but non-aqueous precipitation fouling is also known. Precipitation fouling is a very common problem in boilers and heat exchangers operating with hard water and often results in limescale. Through changes in temperature, or solvent evaporation or degasification, the concentration of salts may exceed the saturation, leading to a precipitation of solids (usually crystals). As an example, the equilibrium between the readily soluble calcium bicarbonate - always prevailing in natural water - and the poorly soluble calcium carbonate, the following chemical equation may be written:

$$Ca(HCO_3)_2(aqueous) \Rightarrow CaCO_3 \downarrow + CO_2 \uparrow + H_2O$$

The calcium carbonate that forms through this reaction precipitates. Due to the temperature dependence of the reaction, and increasing volatility of CO_2 with increasing temperature, the scaling is higher at the hotter outlet of the heat exchanger than at the cooler inlet.

In general, the dependence of the salt solubility on temperature or presence of evaporation will often be the driving force for precipitation fouling. The important distinction is between salts with "normal" or "retrograde" dependence of solubility on temperature. The salts with the "normal" solubility increase their solubility with increasing temperature and thus will foul the cooling surfaces. The salts with "inverse" or "retrograde" solubility will foul the heating surfaces. An example of the temperature dependence of solubility. Calcium sulphate is a common precipitation foulant of heating surfaces due to its retrograde solubility.

Precipitation fouling can also occur in the absence of heating or vaporisation. For example, calcium sulphate decreases it solubility with decreasing pressure. This can lead to precipitation fouling of reservoirs and wells in oil fields, decreasing their productivity with time. Fouling of membranes in reverse osmosis systems can occur due to differential solubility of barium sulphate in solutions of different ionic strength. Similarly, precipitation fouling can occur because of solubility changes induced by other factors, e.g., liquid flashing, liquid degassing, redox potential changes, or mixing of incompatible fluid streams.

The following lists some of the industrially common phases of precipitation fouling deposits observed in practice to form from aqueous solutions:

- Calcium carbonate (calcite, aragonite usually at t > ~50 °C, or rarely vaterite);
- Calcium sulphate (anhydrite, hemihydrate, gypsum);
- Calcium oxalate (e.g., beerstone);
- Barium sulphate (barite);
- Magnesium hydroxide (brucite); magnesium oxide (periclase);

- Silicates (serpentine, acmite, gyrolite, gehlenite, amorphous silica, quartz, cristobalite, pectolite, xonotlite);
- Aluminium oxide hydroxides (boehmite, gibbsite, diaspore, corundum);
- Aluminosilicates (analcite, cancrinite, noselite);
- Copper (metallic copper, cuprite, tenorite);
- Phosphates (hydroxyapatite);
- Magnetite or nickel ferrite ($NiFe_2O_4$) from extremely pure, low-iron water.

The deposition rate by precipitation is often described by the following equations:

Transport: $\frac{dm}{dt} = k_t(C_b - C_i)$

Surface crystallisation: $\frac{dm}{dt} = k_r(C_i - C_e)^{n1}$

Overall: $\frac{dm}{dt} = k_d(C_b - C_e)^{n2}$

where:

m - mass of the material (per unit surface area), kg/m^2

t - time, s

C_b - concentration of the substance in the bulk of the fluid, kg/m^3

C_i - concentration of the substance at the interface, kg/m^3

C_e - equilibrium concentration of the substance at the conditions of the interface, kg/m^3

n1, n2 - order of reaction for the crystallisation reaction and the overall deposition process, respectively, dimensionless

k_t, k_r, k_d - kinetic rate constants for the transport, the surface reaction, and the overall deposition reaction, respectively; with the dimension of m/s (when n1 and n2 = 1)

Particulate Fouling

Fouling by particles suspended in water ("crud") or in gas progresses by a mechanism different than precipitation fouling. This process is usually most important for colloidal particles, i.e., particles smaller than about 1 μm in at least one dimension (but which are much larger than atomic dimensions). Particles are transported to the surface by a number of mechanisms and there they can attach themselves, e.g., by flocculation or coagulation. Note that the attachment of colloidal particles typically involves electrical forces and thus the particle behaviour defies the experience from the macroscopic world. The probability of attachment is sometimes referred to as "sticking probability", P:

$$k_d = Pk_t$$

where k_d and k_t are the kinetic rate constants for deposition and transport, respectively. The value of P for colloidal particles is a function of both the surface chemistry, geometry, and the local thermohydraulic conditions.

An alternative to using the sticking probability is to use a kinetic attachment rate constant, assuming the first order reaction:

$$\frac{dm}{dt} = k_a C_i$$

and then the transport and attachment kinetic coefficients are combined as two processes occurring is series:

$$k_d = \left(\frac{1}{k_a} + \frac{1}{k_t} \right)^{-1}$$

$$\frac{dm}{dt} = k_d C_b$$

where:

- dm/dt is the rate of the deposition by particles, kg m^{-2} s^{-1},
- k_a, k_t and k_d are the kinetic rate constants for deposition, m/s,
- C_i and C_b are the concentration of the particle foulant at the interface and in the bulk fluid, respectively; kg m^3.

Being essentially a surface chemistry phenomenon, this fouling mechanism can be very sensitive to factors that affect colloidal stability, e.g., zeta potential. A maximum fouling rate is usually observed when the fouling particles and the substrate exhibit opposite electrical charge, or near the point of zero charge of either of them.

Particles larger than those of colloidal dimensions may also foul e.g., by sedimentation ("sedimentation fouling") or straining in small-size openings.

With time, the resulting surface deposit may harden through processes collectively known as "deposit consolidation" or, colloquially, "aging".

The common particulate fouling deposits formed from aqueous suspensions include:

- iron oxides and iron oxyhydroxides (magnetite, hematite, lepidocrocite, maghemite, goethite);
- Sedimentation fouling by silt and other relatively coarse suspended matter.

Fouling by particles from gas aerosols is also of industrial significance. The particles can be either solid or liquid. The common examples can be fouling by flue gases, or fouling of air-cooled components by dust in air. The mechanisms are discussed in article on aerosol deposition.

Corrosion Fouling

Corrosion deposits are created in-situ by the corrosion of the substrate. They are distinguished from fouling deposits, which form from material originating ex-situ. Corrosion deposits should not be confused with fouling deposits formed by ex-situ generated corrosion products. Corrosion deposits will normally have composition related to the composition of the substrate. Also, the geometry of the metal-oxide and oxide-fluid interfaces may allow practical distinction between the corrosion and fouling deposits. An example of corrosion fouling can be formation of an iron oxide or oxyhydroxide deposit from corrosion of the carbon steel underneath. Corrosion fouling should not be confused with fouling corrosion, i.e., any of the types of corrosion that may be induced by fouling.

Chemical Reaction Fouling

Chemical reactions may occur on contact of the chemical species in the process fluid with heat transfer surfaces. In such cases, the metallic surface sometimes acts as a catalyst. For example, corrosion and polymerisation occurs in cooling water for the chemical industry which has a minor content of hydrocarbons. Systems in petroleum processing are prone to polymerisation of olefins or deposition of heavy fractions (asphaltenes, waxes, etc.). High tube wall temperatures may lead to carbonising of organic matter. Food industry, for example milk processing, also experiences fouling problems by chemical reactions.

Fouling through an ionic reaction with an evolution of an inorganic solid is commonly classified as precipitation fouling (not chemical reaction fouling).

Solidification Fouling

Solidification fouling occurs when a component of the flowing fluid "freezes" onto a surface forming a solid fouling deposit. Examples may include solidification of wax (with a high melting point) from a hydrocarbon solution, or of molten ash (carried in a furnace exhaust gas) onto a heat exchanger surface. The surface needs to have a temperature below a certain threshold; therefore, it is said to be subcooled in respect to the solidification point of the foulant.

Composite Fouling

Composite fouling is common. This type of fouling involves more than one foulant or more than one fouling mechanism working simultaneously. The multiple foulants or mechanisms may interact with each other resulting in a synergistic fouling which is not a simple arithmetic sum of the individual components.

Fouling on Mars

NASA Mars Exploration Rovers (Spirit and Opportunity) experienced (presumably) abiotic fouling of solar panels by dust particles from the Martian atmosphere. Some of the deposits subsequently spontaneously cleaned off. This illustrates the universal nature of the fouling phenomena.

Quantification of Fouling

The most straight-forward way to quantify fairly uniform fouling is by stating the average deposit surface loading, i.e., kg of deposit per m^2 of surface area. The fouling rate will then be expressed in kg/m^2s, and it is obtained by dividing the deposit surface loading by the effective operating time. The normalised fouling rate (also in kg/m^2s) will additionally account for the concentration of the foulant in the process fluid (kg/kg) during preceding operations, and is useful for comparison of fouling rates between different systems. It is obtained by dividing the fouling rate by the foulant concentration. The fouling rate constant (m/s) can be obtained by dividing the normalised fouling rate by the mass density of the process fluid (kg/m^3). Deposit thickness (μm) and porosity (%) are also often used for description of fouling amount. The relative reduction of diameter of piping or increase of the surface roughness can be of particular interest when the impact of fouling on pressure drop is of interest.

In heat transfer equipment, where the primary concern is often the effect of fouling on heat transfer, fouling can be quantified by the increase of the resistance to the flow of heat (m^2K/W) due to fouling (termed "fouling resistance"), or by development of heat transfer coefficient (W/m^2K) with time. If under-deposit or crevice corrosion is of primary concern, it is important to note non-uniformity of deposit thickness (e.g., deposit waviness), localised fouling, packing of confined regions with deposits, creation of occlusions, "crevices", "deposit tubercles", or sludge piles. Such deposit structures can create environment for underdeposit corrosion of the substrate material, e.g., intergranular attack, pitting, stress corrosion cracking, or localised wastage. Porosity and permeability of the deposits will likely influence the probability of underdeposit corrosion. Deposit composition can also be important - even minor components of the deposits can sometimes cause severe corrosion of the underlying metal (e.g., vanadium in deposits of fired boilers causing hot corrosion).

There is no general rule on how much deposit can be tolerated, it depends on the system. In many cases, a deposit even a few micrometers thick can be troublesome. A deposit in a millimetre-range thickness will be of concern in almost any application.

Progress of Fouling with Time

Deposit on a surface does not always develop steadily with time. The following fouling scenarios can be distinguished, depending on the nature of the system and the local thermohydraulic conditions at the surface:

- *Induction Period:* Sometimes, a near-nil fouling rate is observed when the surface is new or very clean. This is often observed in biofouling and precipitation fouling. After the "induction period", the fouling rate increases.
- *"Negative" Fouling:* This can occur when fouling rate is quantified by monitoring heat transfer. Relatively small amounts of deposit can improve heat transfer, relative to clean surface, and give an appearance of "negative" fouling rate and negative total fouling amount. Negative fouling is often observed under nucleate-boiling heat-transfer conditions (deposit improves bubble nucleation) or forced-convection (if the deposit increases the surface roughness and the surface is no longer "hydraulically smooth"). After the initial period of "surface roughness control", the fouling rate usually becomes strongly positive.
- *Linear Fouling:* The fouling rate can be steady with time. This is a common case.
- *Falling Fouling:* Under this scenario, the fouling rate decreases with time, but never drops to zero. The deposit thickness does not achieves a constant value. The progress of fouling can be often described by two numbers: the initial fouling rate (a tangent to the fouling curve at zero deposit loading or zero time) and the fouling rate after a long period of time (an oblique asymptote to the fouling curve).
- *Asymptotic Fouling:* Here, the fouling rate decreases with time, until it finally reaches zero. At this point, the deposit thickness remains constant with time (a horizontal asymptote). This is often the case for relatively soft or poorly adherent deposits in areas of fast flow. The asymptote is usually interpreted as the deposit loading at which the deposition rate equals the deposit removal rate.
- *Accelerating Fouling:* Under this scenario, the fouling rate increases with time; the rate of deposit buildup accelerates with time (perhaps until it becomes transport limited). Mechanistically, this scenario can develop when fouling increases the surface roughness, or when the deposit surface exhibits higher chemical propensity to fouling than the pure underlying metal.
- *Seesaw Fouling:* Here, fouling loading generally increases with time (often assuming a generally linear or falling rate), but, when looked at in more detail, the fouling progress is periodically interrupted and

takes the form of sawtooth curve. The periodic sharp variations in the apparent fouling amount often correspond to the moments of system shutdowns, startups or other transients in operation. The periodic variations are often interpreted as periodic removal of some of the deposit (perhaps deposit re-suspension due to pressure pulses, spalling due thermal stresses, or exfoliation due to redox transients). Steam blanketing has been postulated to occur between the partially spalled deposits and the heat transfer surface. However, other reasons are possible, e.g., trapping of air inside the surface deposits during shutdowns, or inaccuracy of temperature measurements during transients ("temperature streaming").

Fouling Modelling

Schematics of the fouling process consisting of simultaneous foulant deposition and deposit removal.

Fouling of a system can be modelled as consisting of several steps:

- Generation or ingress of the species that causes fouling ("foulant sourcing");
- Foulant transport with the stream of the process fluid (most often by advection);
- Foulant transport from the bulk of the process fluid to the fouling surface. This transport is often by molecular or turbulent-eddy diffusion, but may also occur by inertial coasting/impaction, particle interception by the surface (for particles with finite sizes), electrophoresis, thermophoresis, diffusiphoresis, Stefan flow (in condensation and evaporation), sedimentation, Magnus force (acting on rotating particles), thermoelectric effect, and other mechanisms.
- Induction period, i.e., a near-nil fouling rate at the initial period of fouling (observed only for some fouling mechanisms);
- Foulant crystallisation on the surface (or attachment of the colloidal particle, or chemical reaction, or bacterial growth);
- Sometimes fouling autoretardation, i.e., reduction (or potentially enhancement) of crystallisation/attachment rate due to changes in the surface conditions caused by the fouling deposit;
- Deposit dissolution (or re-entrainment of loosely attached particles);
- Deposit consolidation on the surface (e.g., through Oswald ripening or differential solubility in temperature gradient) or cementation, which account for deposit losing its porosity and becoming more tenacious with time;
- Deposit spalling, erosion wear, or exfoiliation.

Deposition consists of transport to the surface and subsequent attachment. Deposit removal is either through deposit dissolution, particle re-entrainment, or deposit spalling, erosive wear, or exfoliation. Fouling results from foulant generation, foulant deposition, deposit removal, and deposit consolidation.

The Economic and Environmental Importance of Fouling

Fouling is ubiquitous and generates tremendous operational losses, not unlike corrosion. For example, one estimate puts the losses due to fouling of heat exchangers in industrialised nations to be about 0.25 per cent of their GDP. Another analysis estimated (for 2006) the economical loss due to boiler and turbine fouling in China utilities at 4.68 billion dollars, which is about 0.169 per cent the country GDP. The losses initially result from impaired heat transfer, corrosion damage (in particular under-deposit and crevice corrosion), increased pressure drop, flow blockages, flow redistribution inside components, flow instabilities, induced vibrations (possibly leading to other problems, e.g., fatigue), fretting, premature failure of electrical heating elements, and a large number of other often unanticipated problems. In addition, the ecological costs should be (but typically are not) considered. The ecological costs arise from the use of biocides for the avoidance of biofouling, from the increased fuel input to compensate for the reduced output caused by fouling, and an increased use of cooling water in once-through cooling systems.

For example, "normal" fouling at a conventionally fired 500 MW (net electrical power) power station unit accounts for output losses of the steam turbine of 5 MW and more. In a 1,300 MW nuclear power station, typical losses could be 20 MW and up (up to 100 per cent if the station shuts down due to fouling-induced component degradation). In seawater desalination plants, fouling may reduce the gained output ratio by two-digit percentages (the gained output ratio is an equivalent that puts the mass of generated distillate in relation to the steam used in the process). The extra electrical consumption in compressor-operated coolers is also easily in the two-digit area. In addition to the operational costs, also the capital cost increases because the heat exchangers have to be designed in larger sizes to compensate for the heat-transfer loss due to fouling. To the output losses listed above, one needs to add the cost of down-time required to inspect, clean, and repair the components (millions of dollars per day of shutdown in lost revenue in a typical power plant), and the cost of actually doing this maintenance. Finally, fouling is often a root cause of serious degradation problems that may limit the life of components or entire plants.

Fouling Control

The most fundamental and usually preferred method of controlling fouling is to prevent the ingress of the fouling species into the cooling water circuit.

In steam power stations and other major industrial installations of water technology, macro fouling is avoided by way of pre-filtration and cooling water debris filters. Some plants employ foreign-object exclusion programme (to eliminate the possibility of salient introduction of unwanted materials, e.g., forgetting tools during maintenance). Acoustic monitoring is sometimes employed to monitor for fretting by detached parts.

In the case of micro fouling, water purification is achieved with extensive methods of water treatment, microfiltration, membrane technology (reverse osmosis, electrodeionisation) or ion-exchange resins. The generation of the corrosion products in the water piping systems is often minimised by controlling the pH of the process fluid (typically alkanisation with ammonia, morpholine, ethanolamine or sodium phosphate), control of oxygen dissolved in water (for example, by addition of hydrazine), or addition of corrosion inhibitors.

For water systems at relatively low temperatures, the applied biocides may be classified as follows: inorganic chlorine and bromide compounds, chlorine and bromide cleavers, ozone and oxygen cleavers, unoxidisable biocides. One of the most important unoxidisable biocides is a mixture of chloromethyl-isothiazolinone and methyl-isothiazolinone. Also applied are dibrom nitrilopropionamide and quaternary ammonium compounds. For underwater ship hulls bottom paints are applied.

Chemical fouling inhibitors can reduce fouling in many systems, mainly by interfering with the crystallisation, attachment, or consolidation steps of the fouling process. Examples for water systems are: chelating agents (for example, EDTA), long-chain aliphatic amines or polyamines (for example, octadecylamine, helamin, and other "film-forming" amines), organic phosphonic acids (for example, etidronic acid), or polyelectrolytes (for example, polyacrylic acid, polymethacrylic acid, usually with a molecular weight lower than 10000). For fired boilers, aluminium or magnesium additives can lower the melting point of ash and promote creation of deposits which are easier to remove.

Magnetic water treatment has been a subject of controversy as to its effectiveness for fouling control since 1950s. The prevailing opinion is that it simply "does not work". Nevertheless, some studies suggest that it may be effective under some conditions to reduce buildup of calcium carbonate deposits.

On the component design level, fouling can often (but not always) be minimised by maintaining a relatively high (for example, 2 m/s) and uniform fluid velocity throughout the component. Stagnant regions need to be eliminated. Components are normally overdesigned to accommodate the fouling anticipated between cleanings. However, a significant overdesign can be a design error because it may lead to increased fouling due to reduced velocities. Periodic on-line pressure pulses or backflow can be effective if the capability is carefully incorporated at the design time. Blowdown capability is always incorporated into steam generators or evaporators to control the accumulation of non-volatile impurities that cause or aggravate fouling. Low-fouling surfaces (for example, very smooth, implanted with ions, or of low surface energy like Teflon) are an option for some applications. Modern components are typically required to be designed for ease of inspection of internals and periodic cleaning. On-line fouling monitoring systems are designed for some application so that blowing or cleaning can be applied before unpredictable shutdown is necessary or damage occurs.

Chemical or mechanical cleaning processes for the removal of deposits and scales are recommended when fouling reaches the point of impacting the system performance or an onset of significant fouling-induced degradation (e.g., by corrosion). These processes comprise pickling with acids and complexing agents, cleaning with high-velocity water jets ("water lancing"), recirculating ("blasting") with metal, sponge or other balls, or propelling offline mechanical "bullet-type" tube cleaners. Whereas chemical cleaning causes environmental problems through the handling, application, storage and disposal of chemicals, the mechanical cleaning by means of circulating cleaning balls or offline "bullet-type" cleaning can be a more environmentally-friendly alternative. In some heat-transfer applications, mechanical mitigation with dynamic scraped surface heat exchangers is an option. Also ultrasonic or abrasive cleaning methods are available for many specific applications.

Biofouling

Biofouling or biological fouling is the undesirable accumulation of microorganisms, plants, algae, and/or animals on wetted structures.

Impact

Biofouling is especially economically significant on ships' hulls where high levels of fouling can reduce the performance of the vessel and increase its fuel requirements. Biofouling is also found in almost all circumstances where water based liquids are in contact with other materials. Industrially important examples include membrane systems, such as membrane bioreactors and reverse osmosis spiral wound membranes cooling water cycles of large industrial equipments and power stations. Biofouling can also occur in oil pipelines carrying oils with entrained water especially those carrying used oils, cutting oils, soluble oil or hydraulic oils.

Anti-fouling

Anti-fouling is the process of removing the accumulation, or preventing its accumulation. In industrial processes, bio-dispersants can be used to control biofouling. In less controlled environments, anti-fouling coatings which contain biocides or non-toxic coatings which prevent organisms from attaching can be used.

Biocides

Biocides are chemical substances that can deter or kill the microorganisms responsible for biofouling. Biocides are incorporated into an anti-fouling surface coating, typically physical adsorption or through chemical modification of the surface. Biofouling occurs on surfaces after formation of a biofilm. The biofilm creates a surface onto which successively larger microorganisms can attach. In marine environments this usually concludes with barnacle attachment.

The biocides often target the microorganisms which create the initial biofilm, typically bacteria. Once dead, they are unable to spread and can detach. Other biocides are toxic to larger organisms in biofouling, such as the fungi and algae. The most commonly used biocide, and anti-fouling agent, is the tributyltin moiety (TBT). It is toxic to both microorganisms and larger aquatic organisms. It is estimated that TBT derived anti-fouling coatings cover 70 per cent of the world's vessels.

The prevalence of TBT and other tin based anti-fouling coatings on marine vessels is a current environmental problem. TBT has been shown to cause harm to many marine organisms, specifically oysters and mollusks. Extremely low concentrations of tributyltin moiety (TBT) causes defective shell growth in the oyster *Crassostrea gigas* (at a concentration of 20 ng/l) and development of male characteristics in female genitalia in the dog whelk *Nucella lapillus* (where gonocharacteristic change is initiated at 1 ng/l).

The international maritime community has recognised this problem and there is planned phase out of tin based coatings, including a ban on newly built vessels. This phase out of toxic biocides in marine coatings is a severe problem for the shipping industry; it presents a major challenge for the producers of coatings to develop alternative technologies. Safer methods of biofouling control are actively researched. Copper compounds have successfully been used in paints and continue to be used as metal sheeting (for example Muntz metal which was specifically made for this purpose), though there is still debate as to the safety of copper.

Non-toxic Coatings

Non-toxic anti-fouling coatings prevent any attachment of microorganisms thus negating the use of biocides. Further, these coatings are usually based on polymers and researchers are able to design self-healing coatings.

There are two classes of non-toxic anti-fouling coatings. The most common class relies on low friction and low surface energies. This results in hydrophobic surfaces. These coatings create a smooth surface which can prevent attachment of larger microorganisms. For example, fluoropolymers and silicone coatings are commonly used. These coatings are ecologically inert but have problems with mechanical strength and long term stability. Specifically, after days biofilms (slime) can coat the surfaces which buries the chemical activity and allows microorganisms to attach. The second class of non-toxic anti-fouling coatings are hydrophilic coatings. They rely on high amounts of hydration in order to increase the energetic penalty of removing water for proteins and microorganisms to attach. The most common example of these coatings are based on highly hydrated zwitterions, such as glycine betaine and sulphobetaine. These coatings are also low friction but are superior to hydrophobic surfaces because they prevent bacteria attachment, preventing biofilm formation. These coatings are not yet commercially available and are being designed as part of a larger effort by the Office of Naval Research to develop environmentally safe biomimetic ship coatings.

Types

Biofouling is divided into microfouling — biofilm formation and bacterial adhesion — and macrofouling — attachment of larger organisms, of which the main culprits are barnacles, mussels, polychaete worms, bryozoans, and seaweed. Together, these organisms form a fouling community.

Individually small, accumulated biofoulers can form enormous masses that severely diminish ships' maneuverability and carrying capacity. Fouling causes huge material and economic costs in maintenance of mariculture, shipping industries, naval vessels, and seawater pipelines. Governments and industry spend more than US$ 5.7 billion annually to prevent and control marine biofouling.

History

Bio-fouling, especially of ships, has been a problem for many centuries and probably for as long as mankind has been sailing the oceans. In *Deipnosophistae*, Athenaeus described the anti-fouling efforts taken in the construction of the great ship of Hieron of Syracuse (died 467 BCE). Before the 18th century various graving and paying techniques were used to try and prevent fouling using three main substances: White stuff, which was a mixture of train oil, rosin and brimstone; Black stuff, a mixture of tar and pitch; and Brown stuff, which was simply brimstone added to Black stuff.

Copper sheathing was first suggested in 1708 by Charles Perry, as an anti-fouling device but the first experiments with copper sheathing were not made until the late 1750s, at which time the bottoms and sides of several ships' keels and false keels were sheathed with copper plates.

The copper performed very well in protecting the hull from invasion by worm, and in preventing the growth of weed, for when in contact with water, the copper produced a poisonous film, composed mainly of oxychloride, that deterred these marine creatures. Furthermore, as this film was slightly soluble it gradually washed away, leaving no way in which marine life could attach itself to the ship. From about 1770, the Royal Navy set about coppering the bottoms of the entire fleet and continued to the end of the use of wooden hulled ships. The process was so successful that the term *copper bottomed* came to mean something that was highly dependable or risk free.

Other Areas of Occurrence

Biofouling can also occur in groundwater wells where build-up can limit recovery flow rates, and in the exterior and interior of ocean-laying pipes. In the latter case it has been shown to retard the seawater flow through the pipe and has to be removed with the tube cleaning process. Biofouling also occurs on the surfaces of living marine organisms, when it is known as epibiosis.

Magnetic Water Treatment

Magnetic Water treatment usually by the application of a changing Magnetic flux is a proposed method of water softening The flux change may be deliberately caused by an electromagnet or the result of water flowing near a permanent magnet, usually in a pipe surrounded by the magnetic water treatment transducer device. Magnetic water treatment (also known as Anti-scale Magnetic Treatment or AMT) is a proposed method of reducing the effects of hard water, as an alternative to water softening. Vendors of magnetic water treatment devices have claimed that powerful magnetic fields can affect the structure of water molecules or the properties of solutes passing through the magnetic field, thus eliminating the need for chemical softening agents. Only the effective hardness is claimed to be altered; no solutes (such as calcium or magnesium) are removed from the water by the process.

Most scientific studies do not support these claims and suggest that magnetic water treatment may be ineffective. Certainly many, perhaps most claims for small domestic apparatus may be pseudoscientific because the magnetic field, if it works at all, would have to be very powerful, and this high flux is not easily achieved in a small space.

Mechanism

Duration of exposure and field strength, gradient, rate of change, and orientation along or perpendicular to flow are variously cited as important to the results. Magnetic water treatment proponent Klaus Kronenberg proposed that the shapes of solute lime molecules are modified by strong magnetic fields, leading them to precipitate as spherical or round crystals rather than deposit as sheets or platelets of hard crystals. John Donaldson, professor of chemistry at Brunel University, proposed that the crucial step is the interruption of agglomeration of particles carrying a surface charge after dissolved contaminants have nucleated as a colloidal suspension. Simon Parsons of the School of Water Sciences at Cranfield University proposed that the magnetic field reduces the surface charge on small particles, increasing the tendency to coagulate as large particles that stay with the flow rather than depositing as scale. Some proponents propose that formation of the softer polymorph aragonite over the more common calcite is favoured in the presence of a magnetic field; an internal study in 1996 at Lawrence Livermore National Laboratory found no difference in preferred crystal structure of scale deposited in magnetic water treatment systems.

Liu et al and Coey and Cass published research in 2010 and 2000 demonstrating that magnetic treatment causes water containing minerals to favour formation of more soluble calcium aragonite over calcium carbonate, and the resulting removal of calcium carbonate deposits from a steel substrate. Furthermore, in their 2010 publication, Liu et all conclude that "The magnetic treatment of scaling waters was proved to be efficient. The efficiency obtained with this very simple magnetic device can be very much improved if the geometry is better devised." . Kozic and Lipus concluded in their 2003 paper that the effects of magnetic treatment on water indeed results in reduced formation of limescale and that this effect lasts approximately 200 hours.

Some devices are passive magnets attached to a water pipe, but there is limited scientific explanation as to why these should have any effect. Other devices send a pulsating and reversing electric current through one or more coils located in close proximity around the water pipe, creating pulsating and reversing electro-magnetic fields in the water. Where two or more coils are used, the effect of this is said to be to cause the limescale particles to collide into each other and hence to change their crystal structure (i.e. from calcium carbonate to more soluble calcium aragonite). This effect would appear to depend on the particular properties of the water (e.g. the presence of other elements such as iron and magnesium) and, if it does occur, may be due to other, as yet unidentified, interactions in the water. There are mixed reports from consumers as to the effectiveness of these devices.

Studies of Effectiveness

Scientific and engineering studies generally refute the effectiveness of the method, finding no differences not attributable to other causes between systems with and without a magnetic water treatment device, and no theoretical basis to expect that there might be. Vendors frequently use pictures and testimonials to support their claims, but omit quantitative detail and well-controlled studies. Advertisements and promotions generally omit such system variables as corrosion coupon results or system mass balance analyticals, as well as measurements of post-treatment water such as concentration of hardness ions or the distribution, structure, and morphology of suspended particles.. Nevertheless, some studies do suggest that it may be effective under some conditions to reduce buildup of calcium carbonate deposits.

Other Claimed Effects of Magnetically Treated Water

One study has claimed statistically significant reduction in calculus formation on the teeth when exposed to magnetically treated water. The same study reported that reduction in plaque and gingival index was not statistically significant. If magnetic water treatment were to work, a health benefit, by comparison to conventional water softening treatment using sodium-based ion replacement, would be the avoidance of increasing dietary sodium intake.

Absorption of Water

Water is highly essential for plants for various metabolic activities. Land plants get water supply from soil which serves as the source of water and minerals to them. The way in which water from soil enters roots, particularly to the root xylem, is called *mechanism of water absorption. Both* Active and Passive absorption have been proposed for mechanism of water absorption.

Active Absorption

It is absorption of water by roots with the help of metabolic energy generated by the root respiration. The force for water absorption originates from the cells of root due to root respiration. As the root cells actively take part in the process so it is called Active absorption. According to Renner, active absorption takes place in low transpiring and well-watered plants and 4 per cent of total water absorption is carried out in this process. The active absorption is carried out by two theories which are, Active osmotic water absorption and Active non-osmotic water absorption.

Active Osmotic Water Absorption

This theory was given by Atkins (1916) and Priestley (1921). According to this theory, the root cells behave as ideal osmotic system through which water moves up from soil solution to root xylem along an increasing gradient of D.P.D (suction pressure which is the real force for water absorption). If solute concentration is high and water potential is low in the root cells, water can enter from soil to root cells through endosmosis.

Mineral nutrients are absorbed actively by the root cells due to utilisation of ATP. As a result, the concentration of ions (osmotica) in the xylem vessels is more in comparison to the soil water. A concentration gradient is established between the root and the soil water. Hence, the solute potential of xylem water is more in comparison to that of soil and correspondingly water potential is low than the soil water. Otherwise stated, water potential is comparatively positive in the soil water. This gradient of water potential causes endosmosis. The endosmosis of water continues till the water potential both in the root and soil becomes equal. It is the absorption of minerals that utilise metabolic energy, but not water absorption. Hence, absorption of water is indirectly an active process.

Active Non-osmotic Water Absorption

This theory was given by Thimann (1951) and Kramer (1959). According to the theory, sometimes water is absorbed against concentration gradient. This requires expenditure of metabolic energy released from respiration of root cells. There is no direct evidence, but some scientists suggest involvement of energy from respiration. In conclusion it is said that, the evidences supporting active absorption of water are themselves poor.

Passive Absorption

This mechanism is carried out without utilisation of metabolic energy. Here only the root act as an organ of absorption or passage. Hence, sometimes it is called water absorption 'through roots', rather 'by' roots. It occurs in rapidly transpiring plants during daytime, because of opening of stomata and the atmospheric conditions. The force for absorption of water is created at the leaf end the transpiration pull. The main cause behind this transpiration pull, water is lifted up in the plant axis like a bucket of water is lifted by a person from an well. Transpiration pull is responsible for dragging water at the leaf end, the pull or force is transmitted down to the root through water column in the xylem elements. The continuity of water column remains intact due to the cohesion between the molecules and it act as a rope. Root simply act as a passive organ of absorption. As transpiration proceeds, simultaneously water absorption also takes place to compensate the water loss from leaf end. Most volume of water entering plants is by means of passive absorption.

Conclusion

The water thus absorbed by the plants as described and most of the absorption is by passive absorption. [From:-Parthasarathi Tripathy]

AirDye

AirDye technology manages the application of colour to textiles without the use of water. It was developed and patented by Colorep, a California-based sustainable technology company. The process of making textiles can require several dozen gallons of water for each pound of clothing. The AirDye process employs air instead of water to help the dyes penetrate fibers, a process that uses no water and requires less energy than traditional methods of dyeing, The technology works only on synthetic materials and is currently available only in the United States. Colorep says it plans to extend its use to Europe by the end of summer, and to Central America by late this year.

Technology

The AirDye process employs air instead of water to help the dye penetrate the fibre. AirDye technology heats up fabric, then injects dye directly into the fibers in the form of a gas. The AirDye process uses no water and less energy than traditional methods, while still achieving impressive colours in solids and prints. Airdye is a proprietary technology that cannot be done in the traditional sublimation or heat transfer process. Paper, which can be recycled, continues to be the medium for transferring disperse dyes to fabric.

Applications

AirDye technology works only on synthetic materials. It can be used to dye clothes, swimwear, carpet, curtains, linens, event banners, and ceiling tile.

Characteristics

Environment

The process of making textiles can require several dozen gallons of water for each pound of clothing, especially during the dyeing process. 2.4 trillion gallons of water are used in synthetic dyeing, AirDye's addressable market, each year. AirDye technology eliminates hazardous wastewater as a by-product of dyeing fabric. Water scarcity affects one in three people on every continent and is getting worse as water needs rise with population growth, urbanisation and increased usage by households and industry. Depending on the fabric, and type of dyeing, AirDye uses up to 95 per cent less water, and up to 86 per cent less energy, contributing 84 per cent less to global warming, according to an independent assessment. Additionally, some companies state that with AirDye technology they can print to order which eliminates production excess and waste.

Economics

Colorep claims that its AirDye Process is priced competitively compared to traditional technology. Additionally, Colorep claims that companies using the technology find the AirDye process reduces waste as 10 per cent of traditional, vat-dyed fabric is damaged during the process, whereas with AirDye, only 1 per cent is damaged. Colorep also says that AirDye enables companies to wait longer to decide what colour or print to put onto their fabric, which would reduce the need for apparel makers to guess what colours consumers will want to wear months ahead.

Consumer

The AirDye process manufactures fabric that can be washed at any temperature, with whites or colours, with or without bleach. Because the AirDye process injects the dyes in the fabric and not on the fabric, bleach and cleaning agents do not affect them.

Design

Fabrics dyed with AirDye technology supposedly feel no different than conventionally dyed textiles. They can be dyed all one colour, a different colour on each side, a pattern on one side and a solid colour on the other, or a pattern on both sides.

Customers

Colorep has licensed its AirDye technology to 16 resellers.

Hunter Douglas

Hunter Douglas Hospitality provides window coverings to the hospitality industry. The Company states that using AirDye technology allows it to save resources, money and the environment. Current and future Hunter Douglas Hospitality products featuring AirDye technology include:

Custom-printed window coverings, including roller shades, roman shades, Vignette modern roman shades, and Silhouette window shadings. The AirDye process can apply designs to the shade fabric without interfering with its shading performance or other characteristics. Textile designs that can feature a look on the front surface of the fabric that differs from what is applied to the back. Fabrics can be printed with a different solid colour on each side of the fabric, two different patterns on each side, or combinations of solid colours and patterns.

Apparel

Costello Tagliapietra

Costello Tagliapietra is a New York–based fashion house that launched their Spring 2010 collection using AirDye technology at their fashion show on September 11, 2009. As winners of the Ecco Domani Fashion Foundation Award, they have an eye on creating clothing that can help enhance the global environment. Using AirDye's water saving technology, enables them to pursue that goal, while designing fashionable clothes.

Patagonia

Patagonia is an outdoor clothing company and is a major contributor to several environmental groups and movements. One of the sewing factories in the United States that does contract work for Patagonia is currently installing AirDye.

A Lot To Say

A Lot To Say is an environmentally responsible t-shirt company that has been given an exclusive and unprecedented endorsement from the National Green Energy Council. A Lot To Say t-shirts are coloured and printed with the AirDye process. With AirDye, the Company estimates for every 25,000 t-shirts sold, the planet will save:

* Energy - 1,132,500 megajoules
* Water - 157,500 US gallons (596,000 L)
* Green House Emissions - 57,500 (kg CO_2 equiv. emissions).

Miss Peaches

Miss Peaches is an eco-friendly swimwear company that has implemented an "Eco-Friendly Initiative" where the designers are constantly searching for and applying the most cutting edge developments in printing processes and other innovations that help preserve the environment. Miss Peaches uses the AirDye process to print all of their swimwear and cover-ups.

Julie Apple

Julie Apple is a handbag company that uses AirDye technology for its prints. The Company states that AirDye technology allows it to print to order, eliminating production excess and waste.

Anaerobic Digestion

Anaerobic digestion is a series of processes in which microorganisms break down biodegradable material in the absence of oxygen, used for industrial or domestic purposes to manage waste and/or to release energy.

The digestion process begins with bacterial hydrolysis of the input materials in order to break down insoluble organic polymers such as carbohydrates and make them available for other bacteria. Acidogenic bacteria then convert the sugars and amino acids into carbon dioxide, hydrogen, ammonia, and organic acids.

Acetogenic bacteria then convert these resulting organic acids into acetic acid, along with additional ammonia, hydrogen, and carbon dioxide. Finally, methanogens convert these products to methane and carbon dioxide.

It is used as part of the process to treat biodegradable waste and sewage sludge As part of an integrated waste management system, anaerobic digestion reduces the emission of landfill gas into the atmosphere. Anaerobic digesters can also be fed with purpose-grown energy crops such as maize.

Anaerobic digestion is widely used as a source of renewable energy. The process produces a biogas, consisting of methane and carbon dioxide. This biogas can be used directly as cooking fuel, in combined heat and power gas engines or upgraded to natural gas quality biomethane.

The utilisation of biogas as a fuel helps to replace fossil fuels. The nutrient-rich digestate that is also produced can be used as fertilizer. The technical expertise required to maintain industrial scale anaerobic digesters coupled with high capital costs and low process efficiencies had limited the level of its industrial application as a waste treatment technology.

Anaerobic digestion facilities have, however, been recognised by the United Nations Development Programme as one of the most useful decentralised sources of energy supply, as they are less capital intensive than large power plants.

History

Scientific interest in the manufacturing of gas produced by the natural decomposition of organic matter, was first reported in the seventeenth century by Robert Boyle and Stephen Hale, who noted that flammable gas was released by disturbing the sediment of streams and lakes. In 1808, Sir Humphry Davy determined that methane was present in the gases produced by cattle manure. The first anaerobic digester was built by a leper colony in Bombay, India in 1859. In 1895 the technology was developed in Exeter, England, where a septic tank was used to generate gas for the sewer gas destructor lamp, a type of gas lighting. Also in England, in 1904, the first dual purpose tank for both sedimentation and sludge treatment was installed in Hampton. In 1907, in Germany, a patent was issued for the Imhoff tank, an early form of digester.

Through scientific research, anaerobic digestion gained academic recognition in the 1930s. This research led to the discovery of anaerobic bacteria, the microorganisms that facilitate the process. Further research was carried out to investigate the conditions under which methanogenic bacteria were able to grow and reproduce. This work was developed during World War II, during which in both Germany and France there was an increase in the application of anaerobic digestion for the treatment of manure.

Applications

Anaerobic digestion is particularly suited to organic material and is commonly used for effluent and sewage treatment. Anaerobic digestion is a simple process that can greatly reduce the amount of organic matter, which might otherwise be destined to be dumped at sea, landfilled or burnt in an incinerator.

Almost any organic material can be processed with anaerobic digestion. This includes biodegradable waste materials such as waste paper, grass clippings, leftover food, sewage, and animal waste. The exception to this is woody wastes that are largely unaffected by digestion, as most anaerobes are unable to degrade lignin. The exception being xylophalgeous anaerobes (lignin consumers) or using high temperature pre-treatment such as pyrolisis which breakdown the lignin. Anaerobic digesters can also be fed with specially grown energy crops such as silage for dedicated biogas production. In Germany and continental Europe, these facilities are referred to as *biogas plants*. A *co-digestion* or *co-fermentation* plant is typically an agricultural anaerobic digester that accepts two or more input materials for simultaneous digestion.

In developing countries, simple home and farm-based anaerobic digestion systems offer the potential for cheap, low-cost energy for cooking and lighting. Anaerobic digestion facilities have been recognised by the United Nations Development Programme as one of the most useful decentralised sources of energy supply. From 1975, China and India have both had large government-backed schemes for adaptation of small biogas plants for use in the household for cooking and lighting. At the present time, projects for anaerobic digestion in the developing world can gain financial support through the United Nations Clean Development Mechanism if they are able to show that they provide reduced carbon emissions.

Pressure from environmentally related legislation on solid waste disposal methods in developed countries has increased the application of anaerobic digestion as a process for reducing waste volumes and generating useful by-products. Anaerobic digestion may either be used to process the source separated fraction of municipal waste or alternatively combined with mechanical sorting systems, to process residual mixed municipal waste. These facilities are called mechanical biological treatment plants.

Utilising anaerobic digestion technologies can help to reduce the emission of greenhouse gasses in a number of key ways:

- Replacement of fossil fuels
- Reducing or eliminating the energy footprint of waste treatment plants
- Reducing methane emission from landfills
- Displacing industrially produced chemical fertilizers
- Reducing vehicle movements
- Reducing electrical grid transportation losses.

Methane and power produced in anaerobic digestion facilities can be utilised to replace energy derived from fossil fuels, and hence reduce emissions of greenhouse gases. This is due to the fact that the carbon in biodegradable material is part of a carbon cycle. The carbon released into the atmosphere from the combustion of biogas has been removed by plants in order for them to grow in the recent past. This could have occurred within the last decade, but more typically within the last growing season. If the plants are re-grown, taking the carbon out of the atmosphere once more, the system will be carbon neutral. This contrasts to carbon in fossil fuels that has been sequestered in the earth for many millions of years, the combustion of which increases the overall levels of carbon dioxide in the atmosphere.

If the putrescible waste processed in anaerobic digesters were disposed of in a landfill, it would break down naturally and often anaerobically. In this case, the gas will eventually escape into the atmosphere. As methane is about twenty times more potent as a greenhouse gas than carbon dioxide, this has significant negative environmental effects.

Digester liquor can be used as a fertilizer supplying vital nutrients to soils. The solid, fibrous component of the digested material can be used as a soil conditioner to increase the organic content of soils. The liquor can be used instead of chemical fertilizers that require large amounts of energy to produce and transport. The use of manufactured fertilizers is, therefore, more carbon-intensive than the use of anaerobic digester liquor fertilizer. In countries, such as Spain where there are many organically depleted soils, the markets for the digested solids can be equally as important as the biogas.

In countries that collect household waste, the utilisation of local anaerobic digestion facilities can help to reduce the amount of waste that requires transportation to centralised landfill sites or incineration facilities. This reduced burden on transportation reduces carbon emissions from the collection vehicles. If localised anaerobic digestion facilities are embedded within an electrical distribution network, they can help reduce the electrical losses that are associated with transporting electricity over a national grid.

In Oakland, California at the East Bay Municipal Utility District's (EBMUD) Main Wastewater Treatment Plant (MWWTP), food waste is currently co-digested with primary and secondary municipal wastewater solids and other high-strength wastes. Compared to municipal wastewater solids digestion, food waste digestion has many benefits. Anaerobic digestion of food waste pulp from the EBMUD food waste process provides a higher normalised energy benefit, compared to municipal wastewater solids:

- 730 to 1,300 kWh per dry ton of food waste applied.
- 560 to 940 kWh per dry ton of municipal wastewater solids applied.

Power Generation

Biogas from sewage works is sometimes used to run a gas engine to produce electrical power; some or all of which can be used to run the sewage works. Some waste heat from the engine is then used to heat the digester. It turns out that the waste heat is, in general, enough to heat the digester to the required temperatures.

The power potential from sewage works is limited – in the UK there are about 80 MW total of such generation, with potential to increase to 150 MW, which is insignificant compared to the average power demand in the UK of about 35,000 MW. The scope for biogas generation from non-sewage waste biological matter – energy crops, food waste, abattoir waste, etc. - is much higher, estimated to be capable of about 3,000 MW. Farm biogas plants using animal waste and energy crops are expected to contribute to reducing CO_2 emissions and strengthen the grid while providing UK farmers with additional revenues.

Some countries offer incentives in the form of, for example, Feed-in Tariffs for feeding electricity onto the power grid in order to subsidise green energy production.

Grid Injection

Biogas grid-injection is the injection of biogas into the natural gas grid. As an alternative, the electricity and the heat can be used for on-site generation, resulting in a reduction of losses in the transportation of energy. Typical energy losses in natural gas transmission systems range from 1–2 per cent, whereas the current energy losses on a large electrical system range from 5–8 per cent.

In October 2010, Didcot Sewage Works became the first in the UK to produce biomethane gas supplied to the national grid, for use in up to 200 homes in Oxfordshire.

The Process

There are many microorganisms that are involved in the process of anaerobic digestion including acetic acid-forming bacteria (acetogens) and methane-forming archaea (methanogens). These organisms feed upon the initial feedstock, which undergoes a number of different processes, converting it to intermediate molecules including sugars, hydrogen, and acetic acid, before finally being converted to biogas.

Different species of bacteria are able to survive at different temperature ranges. Ones living optimally at temperatures between 35–40 °C are called mesophiles or mesophilic bacteria. Some of the bacteria can survive at the hotter and more hostile conditions of 55–60 °C; these are called thermophiles or thermophilic bacteria. Methanogens come from the domain of archaea. This family includes species that can grow in the hostile conditions of hydrothermal vents. These species are more resistant to heat and can, therefore, operate at high temperatures, a property that is unique to thermophiles.

As with aerobic systems, the bacteria in anaerobic systems the growing and reproducing microorganisms within them require a source of elemental oxygen to survive. In an anaerobic system, there is an absence of gaseous oxygen. Gaseous oxygen is prevented from entering the system through physical containment in sealed tanks. Anaerobes access oxygen from sources other than the surrounding air.

The oxygen source for these microorganisms can be the organic material itself or may be supplied by inorganic oxides from within the input material. When the oxygen source in an anaerobic system is derived from the organic material itself, the 'intermediate' end-products are primarily alcohols, aldehydes, and organic acids plus carbon dioxide. In the presence of specialised methanogens, the intermediates are converted to the 'final' end-products of methane, carbon dioxide, and trace levels of hydrogen sulphide. In an anaerobic system the majority of the chemical energy contained within the starting material is released by methanogenic bacteria as methane.

Populations of anaerobic microorganisms typically take a significant period of time to establish themselves to be fully effective. Therefore, it is common practice to introduce anaerobic microorganisms from materials with existing populations, a process known as "seeding" the digesters, and typically takes place with the addition of sewage sludge or cattle slurry.

Stages

There are four key biological and chemical stages of anaerobic digestion:

1. Hydrolysis
2. Acidogenesis
3. Acetogenesis
4. Methanogenesis.

In most cases, biomass is made up of large organic polymers. In order for the bacteria in anaerobic digesters to access the energy potential of the material, these chains must first be broken down into their smaller constituent parts. These constituent parts or monomers such as sugars are readily available by other bacteria. The process of breaking these chains and dissolving the smaller molecules into solution is called hydrolysis. Therefore, hydrolysis of these high-molecular-weight polymeric components is the necessary first step in anaerobic digestion. Through hydrolysis the complex organic molecules are broken down into simple sugars, amino acids, and fatty acids.

Acetate and hydrogen produced in the first stages can be used directly by methanogens. Other molecules such as volatile fatty acids (VFAs) with a chain length that is greater than that of acetate must first be catabolised into compounds that can be directly utilised by methanogens.

The biological process of acidogenesis is where there is further breakdown of the remaining components by acidogenic (fermentative) bacteria. Here, VFAs are created along with ammonia, carbon dioxide, and hydrogen sulphide, as well as other by-products. The process of acidogenesis is similar to the way that milk sours.

The third stage of anaerobic digestion is acetogenesis. Here, simple molecules created through the acidogenesis phase are further digested by acetogens to produce largely acetic acid as well as carbon dioxide and hydrogen.

The terminal stage of anaerobic digestion is the biological process of methanogenesis. Here, methanogens utilise the intermediate products of the preceding stages and convert them into methane, carbon dioxide, and water. It is these components that make up the majority of the biogas emitted from the system. Methanogenesis is sensitive to both high and low pHs and occurs between pH 6.5 and pH 8. The remaining, non-digestible material that the microbes cannot feed upon, along with any dead bacterial remains, constitutes the digestate. A simplified generic chemical equation for the overall processes outlined above is as follows:

$$C_6H_{12}O_6 \text{ ® } 3CO_2 + 3CH_4$$

Configuration

Anaerobic digesters can be designed and engineered to operate using a number of different process configurations:

- Batch or continuous
- Temperature: Mesophilic or thermophilic
- Solids content: High solids or low solids
- Complexity: Single stage or multistage.

Batch or Continuous

A batch system is the simplest form of digestion. Biomass is added to the reactor at the start of the process in a batch and is sealed for the duration of the process. Batch reactors suffer from odour issues that can be a severe problem when they are emptied.

In a typical scenario, biogas production will be formed with a normal distribution pattern over time. Operator can use this fact to determine when they believe the process of digestion of the organic matter has completed. As the batch digestion is simple and requires less equipment and lower levels of design work, it is typically a cheaper form of digestion.

In continuous digestion processes, organic matter is constantly added (continuous complete mixed) or added in stages to the reactor (continuous plug flow; first in – first out). Here, the end-products are constantly or periodically removed, resulting in constant production of biogas. A single or multiple digesters in sequence may be used. Examples of this form of anaerobic digestion include continuous stirred-tank reactors (CSTRs), Upflow anaerobic sludge blanket (UASB), Expanded granular sludge bed (EGSB) and Internal circulation reactors (IC).

Temperature

There are two conventional operational temperature levels for anaerobic digesters, which are determined by the species of methanogens in the digesters:

- *Mesophilic*, which takes place optimally around 30-38 °C or at ambient temperatures between 20-45 °C where mesophiles are the primary microorganism present
- *Thermophilic*, which takes place optimally around 49-57 °C at elevated temperatures up to 70 °C where thermophiles are the primary microorganisms present.

A limit case has been reached in Bolivia, with anaerobic digestion in temperature working conditions less than 10 °C. The anaerobic process is very slow, taking more than three times the normal mesophilic time process. In experimental work at University of Alaska Fairbanks, a 1000 litre digester using psychrophiles harvested from "mud from a frozen lake in Alaska" has produced 200–300 litres of methane per day, about 20–30 per cent of the output from digesters in warmer climates.

There are a greater number of species of mesophiles than thermophiles. These bacteria are also more tolerant to changes in environmental conditions than thermophiles. Mesophilic systems are, therefore, considered to be more stable than thermophilic digestion systems. As mentioned above, thermophilic digestion systems are considered to be less stable, the energy input is higher, and more energy is removed from the organic matter. However, the increased temperatures facilitate faster reaction rates and, hence, faster gas yields. Operation at higher temperatures facilitates greater sterilisation of the end-digestate. In countries where legislation, such as the Animal By-Products Regulations in the European Union, requires end products to meet certain levels of reduction in the amount of bacteria in the output material, this may be a benefit.

Certain processes shred the waste finely and use a short high-temperature and pressure pre-treatment (pasteurisation / hygienisation) stage that significantly enhances the gas output of the following standard mesophilic stage. The hygienisation process is also applied in order to reduce the pathogenic micro-organisms in the feedstock. Hygienisation/pasteurisation may be achieved by using a Landia BioChop hygienisation unit or similar method of combined heat treatment and solids maceration.

A drawback of operating at thermophilic temperatures is that more heat energy input is required to achieve the correct operational temperatures. This increase in energy may not be outweighed by the increase in the outputs of biogas from the systems. Therefore, it is important to consider an energy balance for these systems.

Solids

In a typical scenario, there are three different operational parameters associated with the solids content of the feedstock to the digesters:

- High-solids (dry—stackable substrate)
- High-solids (wet—pumpable substrate)
- Low-solids (wet—pumpable substrate).

High-solids (dry) digesters are designed to process materials with a high-solids content between ~25-40 per cent. Unlike wet digesters that process pumpable slurries, high solids (dry – stackable substrate) digesters are designed to process solid substrates without the addition of water. There are three primary styles of dry digesters: continuous vertical plug flow and batch tunnel horizontal digesters. Continuous vertical plug flow are upright, cylindrical tanks where feedstock is continuously fed to the top of the digester and flows downward by gravity during digestion. In batch tunnel digesters, the feedstock is deposited in tunnel-like chambers with a gas-tight door. Neither approach has mixing inside the digester. The amount of pretreatment such as contaminant removal depends both upon the nature of the waste streams being processed and the desired quality of the digestate. Grinding for size reduction is beneficial in continuous vertical systems as it accelerates digestion, while batch systems avoid grinding and instead require structure (e.g. yard waste) to reduce compaction of the stacked pile. Continuous vertical dry digesters have a smaller footprint due to the shorter effective retention time and vertical design.

Wet digesters can be designed to operate in either a high-solids content, with a total suspended solids (TSS) concentration greater than ~20 per cent, or a low-solids concentration less than ~15 per cent.

High-solids (wet) digesters process a thick slurry that requires more energy input to move and process the feedstock. The thickness of the material may also lead to associated problems with abrasion. High-solids digesters will typically have a lower land requirement due to the lower volumes associated with the moisture. High solids digesters require correction of conventional performance calculations (e.g. gas production, retention time, kinetics, etc.) originally based on very dilute sewage digestion concepts, since larger fractions of the feedstock mass are potentially convertible to biogas.

Low-solids (wet) digesters can transport material through the system using standard pumps that require significantly lower energy input. Low-solids digesters require a larger amount of land than high-solids due to the increase volumes associated with the increased liquid-to-feedstock ratio of the digesters. There are benefits associated with operation in a liquid environment, as it enables more thorough circulation of materials and contact between the bacteria and their food. This enables the bacteria to more readily access the substances they are feeding off and increases the speed of gas yields.

Number of Stages

Digestion systems can be configured with different levels of complexity:

- One-stage or single-stage
- Two-stage or multistage.

A single-stage digestion system is one in which all of the biological reactions occur within a single sealed reactor or holding tank. Utilising a single stage reduces construction costs, however facilitates less control of the reactions occurring within the system. Acidogenic bacteria, through the production of acids, reduce the pH of the tank. Methanogenic bacteria, as outlined earlier, operate in a strictly defined pH range. Therefore, the biological reactions of the different species in a single-stage reactor can be in direct competition with each other. Another one-stage reaction system is an anaerobic lagoon. These lagoons are pond-like earthen basins used for the treatment and long-term storage of manures. Here the anaerobic reactions are contained within the natural anaerobic sludge contained in the pool.

In a two-stage or multi-stage digestion system, different digestion vessels are optimised to bring maximum control over the bacterial communities living within the digesters. Acidogenic bacteria produce organic acids and more quickly grow and reproduce than methanogenic bacteria. Methanogenic bacteria require stable pH and temperature in order to optimise their performance.

Under typical circumstances, hydrolysis, acetogenesis, and acidogenesis occur within the first reaction vessel. The organic material is then heated to the required operational temperature (either mesophilic or thermophilic) prior to being pumped into a methanogenic reactor. The initial hydrolysis or acidogenesis tanks prior to the methanogenic reactor can provide a buffer to the rate at which feedstock is added. Some European countries require a degree of elevated heat treatment in order to kill harmful bacteria in the input waste. In this instance, there may be a pasteurisation or sterilisation stage prior to digestion or between the two digestion tanks. It should be noted that it is not possible to completely isolate the different reaction phases, and often there is some biogas that is produced in the hydrolysis or acidogenesis tanks.

Residence

The residence time in a digester varies with the amount and type of feed material, the configuration of the digestion system, and whether it be one-stage or two-stage.

In the case of single-stage thermophilic digestion, residence times may be in the region of 14 days, which, compared to mesophilic digestion, is relatively fast. The plug-flow nature of some of these systems will mean that the full degradation of the material may not have been realised in this timescale. In this event, digestate exiting the system will be darker in colour and will typically have more odour.

In two-stage mesophilic digestion, residence time may vary between 15 and 40 days.

In the case of mesophilic UASB digestion, hydraulic residence times can be (1 hour to 1 day) and solid retention times can be up to 90 days. In this manner, the UASB system is able to separate solid an hydraulic retention times with the utilisation of a sludge blanket.

Continuous digesters have mechanical or hydraulic devices, depending on the level of solids in the material, to mix the contents enabling the bacteria and the food to be in contact. They also allow excess material to be continuously extracted to maintain a reasonably constant volume within the digestion tanks.

Feedstocks

The most important initial issue when considering the application of anaerobic digestion systems is the feedstock to the process. Digesters typically can accept any biodegradable material; however, if biogas production is the aim, the level of putrescibility is the key factor in its successful application. The more putrescible (digestible) the material the higher the gas yields possible from the system.

Substrate composition is a major factor in determining the methane yield and methane production rates from the digestion of biomass. Techniques to determine the compositional characteristics of the feedstock are available, while parameters such as solids, elemental, and organic analyses are important for digester design and operation.

Anaerobes can break down material with varying degrees of success from readily, in the case of short-chain hydrocarbons such as sugars, to over longer periods of time, in the case of cellulose and hemicellulose. Anaerobic microorganisms are unable to break down long-chain woody molecules such as lignin. Anaerobic digesters were originally designed for operation using sewage sludge and manures. Sewage and manure are not, however, the material with the most potential for anaerobic digestion, as the biodegradable material has already had much of the energy content taken out by the animal that produced it. Therefore, many digesters operate with *co-digestion* of two or more types of feedstock. For example, in a farm-based digester that uses dairy manure as the primary feedstock, the gas production may be significantly increased by adding a second feedstock, e.g., *grass* and *corn* (typical on-site feedstock), or various organic byproducts, such as *slaughterhouse waste*, *fats oils and grease* from restaurants, *organic household waste*, etc. (typical off-site feedstock)..

Digestors processing dedicated energy crops can achieve high levels of degradation and biogas production. Slurry-only systems are generally cheaper but generate far less energy than those using crops such as maize and grass silage; by using a modest amount of crop material (30 per cent), an AD plant can increase energy output tenfold for only three times the capital cost, relative to a slurry-only system.

A second consideration related to the feedstock is moisture content. Dryer, stackable substrates, such as food- and yard-waste, are suitable for digestion in tunnel-like chambers. Tunnel-style systems typically have near-zero wastewater discharge as well, so this style of system has advantages where the discharge of digester liquids are a liability. The wetter the material the more suitable it will be to handling with standard pumps instead of energy intensive concrete pumps and physical means of movement. Also the wetter the material the more volume and area it takes up relative to the levels of gas that are produced. The moisture content of the target feedstock will also affect what type of system is applied to its treatment. In order to use a high-solids anaerobic digester for dilute feedstocks, bulking agents such as compost should be applied to increase the solid content of the input material. Another key consideration is the carbon:nitrogen ratio of the input material. This ratio is the balance of food a microbe requires in order to grow. The optimal C:N ratio for the 'food' a microbe is 20–30:1. Excess N can lead to ammonia inhibition of digestion.

The level of contamination of the feedstock material is a key consideration. If the feedstock to the digesters has significant levels of physical contaminants such as plastic, glass, or metals, then pre-processing will be required in order for the material to be used. If it is not removed then the digesters can be blocked and will not function efficiently. It is with this understanding that mechanical biological treatment plants are designed. The higher the level of pre-treatment a feedstock requires the more processing machinery will be required, and, hence, the project will have higher capital costs.

After sorting or screening to remove any physical contaminants, such as metals, and plastics from the feedstock, the material is often shredded, minced, and mechanically or hydraulically pulped to increase the surface area available to microbes in the digesters and, hence, increase the speed of digestion. The maceration of solids can be achieved by using a chopper pump to transfer the feedstock material into the airtight digester, where anaerobic treatment takes place.

Products

There are three principal products of anaerobic digestion: biogas, digestate, and water.

Biogas

Typical composition of biogas:

Matter	%
Methane, CH_4	50–75
Carbon dioxide, CO_2	25–50
Nitrogen, N_2	0–10
Hydrogen, H_2	0–1
Hydrogen sulphide, H_2S	0–3
Oxygen, O_2	0–2

7t of the bacteria feeding off the input biodegradable feedstock (the methanogenesis stage of anaerobic digestion is performed by archaea - a micro-organism on a distinctly different branch of the phylogenetic tree of life to bacteria), and is mostly methane and carbon dioxide, with a small amount hydrogen and trace hydrogen sulphide. (As-produced, biogas also contains water vapour, with the fractional water vapour volume a function of biogas temperature). Most of the biogas is produced during the middle of the digestion, after the bacterial population has grown, and tapers off as the putrescible material is exhausted. The gas is normally stored on top of the digester in an inflatable gas bubble or extracted and stored next to the facility in a gas holder.

The methane in biogas can be burned to produce both heat and electricity, usually with a reciprocating engine or microturbine often in a cogeneration arrangement where the electricity and waste heat generated are used to warm the digesters or to heat buildings. Excess electricity can be sold to suppliers or put into the local grid. Electricity produced by anaerobic digesters is considered to be renewable energy and may attract subsidies. Biogas does not contribute to increasing atmospheric carbon dioxide concentrations because the gas is not released directly into the atmosphere and the carbon dioxide comes from an organic source with a short carbon cycle.

Biogas may require treatment or 'scrubbing' to refine it for use as a fuel. Hydrogen sulphide is a toxic product formed from sulphates in the feedstock and is released as a trace component of the biogas.

National environmental enforcement agencies such as the U.S. Environmental Protection Agency or the English and Welsh Environment Agency put strict limits on the levels of gasses containing hydrogen sulphide, and, if the levels of hydrogen sulphide in the gas are high, gas scrubbing and cleaning equipment (such as amine gas treating) will be needed to process the biogas to within regionally accepted levels. An alternative method to this is by the addition of ferrous chloride $FeCl_2$ to the digestion tanks in order to inhibit hydrogen sulphide production.

Volatile siloxanes can also contaminate the biogas; such compounds are frequently found in household waste and wastewater. In digestion facilities accepting these materials as a component of the feedstock, low-molecular-weight siloxanes volatilise into biogas. When this gas is combusted in a gas engine, turbine, or boiler, siloxanes are converted into silicon dioxide (SiO_2), which deposits internally in the machine, increasing wear and tear. Practical and cost-effective technologies to remove siloxanes and other biogas contaminants are available at the present time. In certain applications, *in situ* treatment can be used to increase the methane purity by reducing the offgas carbon dioxide content, purging the majority of it in a secondary reactor.

In countries such as Switzerland, Germany, and Sweden, the methane in the biogas may be concentrated in order for it to be used as a vehicle transportation fuel or input directly into the gas mains. In countries where the driver for the utilisation of anaerobic digestion are renewable electricity subsidies, this route of treatment is less likely, as energy is required in this processing stage and reduces the overall levels available to sell.

Digestate

Digestate is the solid remnants of the original input material to the digesters that the microbes cannot use. It also consists of the mineralised remains of the dead bacteria from within the digesters. Digestate can come in three forms: fibrous, liquor, or a sludge-based combination of the two fractions. In two-stage systems, the different forms of digestate come from different digestion tanks. In single-stage digestion systems, the two fractions will be combined and, if desired, separated by further processing.

The second by-product (acidogenic digestate) is a stable organic material consisting largely of lignin and cellulose, but also of a variety of mineral components in a matrix of dead bacterial cells; some plastic may be present. The material resembles domestic compost and can be used as compost or to make low-grade building products such as fibreboard. The solid digestate can also be utilised as feedstock for ethanol production.

The third by-product is a liquid (methanogenic digestate) that is rich in nutrients and can be used as a fertilizer dependent on the quality of the material being digested. Levels of potentially toxic elements (PTEs) should be chemically assessed.

This will be dependent upon the quality of the original feedstock. In the case of most clean and source-separated biodegradable waste streams, the levels of PTEs will be low. In the case of wastes originating from industry, the levels of PTEs may be higher and will need to be taken into consideration when determining a suitable end use for the material.

Digestate typically contains elements such as lignin that cannot be broken down by the anaerobic microorganisms. Also the digestate may contain ammonia that is phytotoxic and will hamper the growth of plants if it is used as a soil-improving material.

For these two reasons, a maturation or composting stage may be employed after digestion. Lignin and other materials are available for degradation by aerobic microorganisms such as fungi, helping reduce the overall volume of the material for transport. During this maturation, the ammonia will be broken down into nitrates, improving the fertility of the material and making it more suitable as a soil improver. Large composting stages are typically used by dry anaerobic digestion technologies.

Wastewater

The final output from anaerobic digestion systems is water. This water originates both from the moisture content of the original waste that was treated but also includes water produced during the microbial reactions in the digestion systems. This water may be released from the dewatering of the digestate or may be implicitly separate from the digestate. The wastewater exiting the anaerobic digestion facility will typically have elevated levels of biochemical oxygen demand (BOD) and chemical oxygen demand (COD). These are measures of the reactivity of the effluent and show an ability to pollute. Some of this material is termed 'hard COD', meaning that it cannot be accessed by the anaerobic bacteria for conversion into biogas. If this effluent were put directly into watercourses, it would negatively affect them by causing eutrophication. As such, further treatment of the wastewater is often required. This treatment will typically be an oxidation stage wherein air is passed through the water in a sequencing batch reactors or reverse osmosis unit.

Aqua Sciences

Aqua Sciences is a Miami Beach-based company providing advanced water technologies. Without any direct funding, they created a module capable of extracting up to 2500 gallons of water from the moisture present in the air.

The Module

They call it the "40-foot Emergency Water Station." It is basically a modified 40-foot trailer that permits the extraction of water from the moisture in the air. It has options such as additional storage tanks for keeping the water for extended periods of time. Also, it can be powered by an internal diesel generator for a week without needing to refuel, or just plugged into the electrical grid present in the area to suffice to its power needs. The module can be transported literally anywhere, used anywhere and under any conditions, as it is essentially the trailer of an 18-wheeler. Another advantage is its ease of use and minimal maintenance needs. It is also possible to add a reverse osmosis module which increases production up to 8000 gallons a day.

Advantages

The advantages of the technology are important such as no toxic or harmful byproducts and it is environmentally friendly. The only requirement for it is 14 per cent humidity in the air, so it can be used in deserts. The water provided is also very pure.

Applications

The United States Army has shown interest in the project, mainly because of the high cost of water transportation to its expeditionary forces. Using the Aqua Sciences' module, that price is pushed down to 0.15$ USD per gallon, which would provide huge logistic savings for the military. Also, it would be practical for providing water after a natural disaster such as the 2004 Indian Ocean earthquake or Hurricane Katrina. It might also provide a solution to the worldwide hydration problem.

Bubble Curtain

A bubble curtain is a system that produces bubbles in a deliberate arrangement in water. The technique is based on bubbles of air (gas) being let out under the water surface, commonly on the bottom. When the bubbles rise they act as a barrier, a curtain, breaking the propagation of waves or the spreading of particles and other contaminants.

Uses

It can be used for the following purposes:

- to reduce propagation of shock waves (eg. acoustic waves from engines or pile driving, explosions etcetera),
- to reduce liquid or debris floating on the surface from spreading
- to control the movements of fish
- for decoration and airing in aquariums.

In June 2010, Okaloosa County, Florida used air bubble curtains to help protect their Destin Pass coastline from oil produced in the Gulf of Mexico by the Deepwater Horizon oil spill. They hoped to push oil up to the surface for booms and skimming boats to collect the oil. British oil company BP, who the U.S. government named as the responsible party for the oil spill, was billed for the cost of the project.

Equipment

The technical system basically consists of a compressor and pipe or hose with nozzles. When used to reduce acoustic waves from pile driving, a distribution manifold made of plastic or rubber is commonly used.

Carbon Filtering

Carbon filtering is a method of filtering that uses a piece of activated carbon to remove contaminants and impurities, utilising chemical adsorption.

Each piece of carbon is designed to provide a large section of surface area, in order to allow contaminants the most possible exposure to the filter media. One pound (454g) of activated carbon contains a surface area of approximately 100 acres.

This carbon is generally activated with a positive charge and is designed to attract negatively charged water contaminants. Carbon filtering is commonly used for water purification, but is also used in air purifiers.

Carbon filters are most effective at removing chlorine, sediment, and volatile organic compounds (VOCs) from water. They are not effective at removing minerals, salts, and dissolved inorganic compounds. Typical particle sizes that can be removed by carbon filters range from 0.5 to 50 micrometers. The particle size will be used as part of the filter description. The efficacy of a carbon filter is also based upon the flow rate regulation. When the water is allowed to flow through the filter at a slower rate, the contaminants are exposed to the filter media for a longer amount of time.

Types of Carbon Filters

There are two predominant types of carbon filters used in the filtration industry: powdered block filters and granular activated filters. In general, carbon block filters are more effective at removing a larger number of contaminants, based upon the increased surface area of carbon. Many carbon filters also use secondary media, such as silver or KDF-55, to prevent bacteria growth within the filter.

History of Carbon Filters

Carbon filters have been used for several hundred years and are considered one of the oldest means of water purification. Historians have shown evidence that carbon filtration may have been used in ancient Egyptian cultures for both air and water sanitisation. 2000 B.C. Sanskrit text refers to filtering water through charcoal (1905 translation of "Sushruta Samhita" by Francis Evelyn Place). The first modern use of a carbon filter to purify potable water occurred in 1862. Carbon filtration was further advanced in the mid-1970s by H. Allen Rice and Alvin E. Rice when they first manufactured a porous carbon block for drinking water use.

Currently, carbon filters are used in individual homes as point-of-use water filters and, occasionally, in municipal water treatment facilities. They are also used as pre-treatment devices for reverse osmosis systems and as specialised filters designed to remove chlorine-resistant cysts, such as giardia and cryptosporidium.

Hydrogen Production

For small scale production of hydrogen water purifiers are installed to prevent formation of minerals on the surface of the electrodes and to remove organics and chlorine from utility water. First the water passes through a 20 micrometer interference (mesh or screen filter) filter to remove sand and dust particles, second, a charcoal filter (activated carbon) to remove organics and chlorine, third stage, a de-ionising filter to remove metallic ions. A test can be done before and after the filter for proper functioning on barium, calcium, potassium, magnesium, sodium and silicon.

Radiation or Nuclear Medicine

Carbon filters, along with HEPA filters, are widely used in the construction of hot cells. This allows the room to exhaust air that contains infinitesimal quantities of radioactivity and contaminants.

Coarse Bubble Diffusers

Coarse bubble diffusers are a pollution control technology used to aerate and or mix wastewater for sewage treatment. They produce 1/4 to 1/2 inch bubbles which rise rapidly from the floor of a wastewater treatment plant or sewage treatment plant tank. They are typically used in grit chambers, equalisation basins, chlorine contact tanks, and aerobic digesters, and sometimes also in aeration tanks. Generally they are better at vertically "pumping" water than at mass transfer of oxygen. Coarse bubble diffusers typically provide half the mass transfer of oxygen as compared to fine bubble diffusers, given the same air volume.

Application

Often in non-newtonian/pseudoplastic fluids, such as a digester with high solids concentration, it does make sense to use coarse bubble diffusers rather than fine bubble diffusers, due to the larger bubbles' ability to shear through more viscous wastewater.

However, over the past two decades, coarse bubble diffusers have been used less frequently, primarily due to the ever increasing cost of energy and the availability of more reliable, highly efficient fine bubble diffusers. Manufacturers of diffused aeration systems claim that converting from coarse bubble to fine bubble system should yield a 50 per cent energy cost savings. Specifically, in aeration tanks, a system that utilises coarse bubble diffusers requires 30 to 40 per cent percent more process air than a fine bubble diffused air system to provide the same level of treatment.

The exception would be in secondary treatment (or side processing) phases. In these processing tanks, floc particles, sediment and carbonate buildup tend to plug or clog the small air release openings on the fine bubble diffusers. Because of their small air openings, fine bubble diffusers cease to have an advantage. Currently, coarse bubble diffusers are the mainstay solution.

These diffusers are typically made in the shape of a perforated rectangular pipe called a wide band, or a cap of 3" in diameter with an elastomeric membrane. Other varieties of coarse bubble diffusers exist, though it is generally accepted that all of them perform similarly with respect to mass oxygen transfer. When comparing disc-shaped diffusers, the majority fail to withstand specific challenges, beyond 1 or 2 years, which include: clogging, blowing off and cracking. Any coarse bubble diffuser that eliminates these problems would deliver a huge cost-savings, not only in product replacement, but in system downtime to facilitate their exchange. Definitely a motivating factor of budget-sensitive operators at municipal waste water treatment processing plants.

Diffuser (Sewage)

An air diffuser or membrane diffuser is an aeration device typically in the shape of a disc, tube or plate, which is used to transfer air and with that oxygen into sewage or industrial wastewater. Oxygen is required by microorganisms/ bacteria residents in the water to break down the pollutants. Diffusers use either rubber membranes or ceramic elements typically and produce either fine or coarse bubbles.

Types of Diffusers

Diffusers are generally referred to as either:

1. Fine Bubble/Fine Pore
2. Coarse Bubble.

Other diffused aeration devices include: jet aerators, aspirators, and U tubes.

Design Specifications

Typical efficiency of a full floor coverage diffused aeration system in clean water is 2%/ft submergence or 6.6%/m submergence. When converted to mass transfer into process or dirty water, it is typically closer to about half of those figures. Manufacturers of fine bubble systems have supported claims that the type, number and size of "pores" have a great effect on efficiency of a diffused aeration system.

Types of Diffused Aeration Systems

Diffusers are typically connected to a piping system which is supplied with pressurised air by a blower. This system is commonly referred to as a diffused aeration system or aeration grid There are two main types of diffused aeration systems, retrievable and fixed grid, that are designed to serve different purposes. In the case of a plant with a single tank, a retrievable system is desirable, in order to avoid stopping operation of the plant when maintenance is required on the aeration system. Fixed systems, on the other hand, are typically less costly, and often more efficient because it is easier to make full use of the floor.

Automated software is available on the web to assist with drafting of aeration systems in CAD, as well as calculation software to help determine diffuser requirements for a given wastewater.

Developments in Membrane Coatings

A developments in recent years has been surface coatings of PTFE on EPDM membranes. This provides a buffer between the EPDM substrate and wastewater, hence reducing the likelihood of chemical attack and oxidation, and also providing better resistance to biological fouling and calcium scaling (manufacturers claim).

Dishwasher Salt

Dishwasher salt is a particular grade of crystalline sodium chloride intended for use in dishwashers. Dishwasher salt is used for the same purpose as (and is very similar to) water softener salt, namely to regenerate ion exchange resins used to remove the hardness ions calcium and magnesium from water. Dishwasher salt is in a granular form, with particles larger than common table salt and unlike table salt has no added iodine (included as a dietary supplement in the U.S.). The granule size ensures that the salt dissolves slowly and that fine particles do not block the softener unit.

Ingredients

95 per cent sodium chloride (salt), 5 per cent other (Anticaking agent e.g. sodium ferrocyanide or potassium ferrocyanide)

Electrodialysis

Electrodialysis (ED) is used to transport salt ions from one solution through ion-exchange membranes to another solution under the influence of an applied electric potential difference. This is done in a configuration called an electrodialysis cell. The cell consists of a feed (diluate) compartment and a concentrate (brine) compartment formed by an anion exchange membrane and a cation exchange membrane placed between two electrodes. In almost all practical electrodialysis processes, multiple electrodialysis cells are arranged into a configuration called an electrodialysis stack, with alternating anion and cation exchange membranes forming the multiple electrodialysis cells. Electrodialysis processes are different compared to distillation techniques and other membrane based processes (such as reverse osmosis) in that dissolved species are moved away from the feed stream rather than the reverse. Because the quantity of dissolved species in the feed stream is far less than that of the fluid, electrodialysis offers the practical advantage of much higher feed recovery in many applications.

Method

The E stream is the electrode stream that flows past each electrode in the stack. This stream may consist of the same composition as the feed stream (e.g., sodium chloride) or may be a separate solution containing a different species (e.g., sodium sulphate). Depending on the stack configuration, anions and cations from the electrode stream may be transported into the C stream, or anions and cations from the D stream may be transported into the E stream. In each case, this transport is necessary to carry current across the stack and maintain electrically neutral stack solutions.

Anode and Cathode Reactions

Reactions take place at each electrode. At the cathode,

$$2e^- + 2\ H_2O \rightarrow H_2\ (g) + 2\ OH^-$$

while at the anode,

$$H_2O \rightarrow 2\ H^+ + \tfrac{1}{2}\ O_2\ (g) + 2e^- \text{ or } 2\ Cl^- \rightarrow Cl_2\ (g) + 2e^-$$

Small amounts of hydrogen gas are generated at the cathode and small amounts of either oxygen or chlorine gas (depending on composition of the E stream and end ion exchange membrane arrangement) at the anode. These gases are typically subsequently dissipated as the E stream effluent from each electrode compartment is combined to maintain a neutral pH and discharged or re-circulated to a separate E tank. However, some (e.g.,) have proposed collection of hydrogen gas for use in energy production.

Efficiency

Current efficiency is a measure of how effective ions are transported across the ion exchange membranes for a given applied current. Typically current efficiencies >80 per cent are desirable in commercial stacks to minimise energy operating costs. Low current efficiencies indicate water splitting in the diluate or concentrate streams, shunt currents between the electrodes, or back-diffusion of ions from the concentrate to the diluate could be occurring.

Current efficiency is calculated according to:

$$\xi = \frac{zFQ_f(C^d_{inlet} - C^d_{outlet})}{NI}$$

where

z = current utilisation efficiency

z = charge of the ion

F = Faraday constant, 96,485 Amp-s/mol

Q_f= diluate flow rate, L/s

C^d_{inlet} = diluate ED cell inlet concentration, mol/L

C^d_{outlet} = diluate ED cell outlet concentration, mol/L

N = number of cell pairs

I = current, Amps.

Current efficiency is generally a function of feed concentration.

Applications

In application, electrodialysis systems can be operated as continuous production or batch production processes. In a continuous process, feed is passed through a sufficient number of stacks placed in series to produce the final desired product quality. In batch processes, the diluate and/or concentrate streams are re-circulated through the electrodialysis systems until the final product or concentrate quality is achieved.

Electrodialysis is usually applied to deionisation of aqueous solutions. However, desalting of sparingly conductive aqueous organic and organic solutions is also possible. Some applications of electrodialysis include:

- Large scale brackish and seawater desalination and salt production.
- Small and medium scale drinking water production (e.g., towns & villages, construction & military camps, nitrate reduction, hotels & hospitals)
- Water reuse (e.g., industrial laundry wastewater, produced water from oil/gas production, cooling tower make-up & blowdown, metals industry fluids, wash-rack water)
- Pre-demineralisation (e.g., boiler make-up & pretreatment, ultrapure water pretreatment, process water desalination, power generation, semiconductor, chemical manufacturing, food and beverage)
- Food processing
- Agricultural water (e.g., water for greenhouses, hydroponics, irrigation, livestock)
- Glycol desalting (e.g., antifreeze / engine-coolants, capacitor electrolyte fluids, oil and gas dehydration, conditioning and processing solutions, industrial heat transfer fluids, secondary coolants from heating, venting, and air conditioning (HVAC))
- Glycerin Purification.

The major application of electrodialysis has historically been the desalination of brackish water or seawater as an alternative to RO for potable water production and seawater concentration for salt production (primarily in Japan). In normal potable water production without the requirement of high recoveries, RO (Reverse Osmosis) is generally believed to be more cost-effective when total dissolved solids (TDS) are 3,000 parts per million (ppm) or greater, while electrodialysis is more cost-effective for TDS feed concentrations less than 3,000 ppm or when high recoveries of the feed are required.

Another important application for electrodialysis is the production of pure water and ultrapure water by electrodeionisation (EDI). In EDI, the purifying compartments and sometimes the concentrating compartments of the electrodialysis stack are filled with ion exchange resin. When fed with low TDS feed (e.g., feed purified by RO), the product can reach very high purity levels (e.g., 18 MΩ-cm). The ion exchange resins act to retain the ions, allowing these to be transported across the ion exchange membranes. The main usage of EDI systems are in electronics, pharmaceutical, power generation, and cooling tower applications.

Limitations

Electrodialysis has inherent limitations, working best at removing low molecular weight ionic components from a feed stream. Non-charged, higher molecular weight, and less mobile ionic species will not typically be significantly removed. Also, in contrast to RO, electrodialysis becomes less economical when extremely low salt concentrations in the product are required and with sparingly conductive feeds: current density becomes limited and current utilisation efficiency typically decreases as the feed salt concentration becomes lower, and with fewer ions in solution to carry current, both ion transport and energy efficiency greatly declines.

Consequently, comparatively large membrane areas are required to satisfy capacity requirements for low concentration (and sparingly conductive) feed solutions. Innovative systems overcoming the inherent limitations of electrodialysis (and RO) are available; these integrated systems work synergistically, with each sub-system operating in its optimal range, providing the least overall operating and capital costs for a particular application.

As with RO, electrodialysis systems require feed pretreatment to remove species that coat, precipitate onto, or otherwise "foul" the surface of the ion exchange membranes. This fouling decreases the efficiency of the electrodialysis system. Species of concern include calcium and magnesium hardness, suspended solids, silica, and organic compounds. Water softening can be used to remove hardness, and micrometer or multimedia filtration can be used to remove suspended solids. Hardness in particular is a concern since scaling can build up on the membranes. Various chemicals are also available to help prevent scaling. Also, electrodialysis reversal systems seek to minimise scaling by periodically reversing the flows of diluate and concentrate and polarity of the electrodes.

Filter Cake

A filter cake is formed by the substances that are retained on a filter. The filter cake grows in the course of filtration, becomes "thicker" as particulate matter is being retained. With increasing layer thickness the flow resistance of the filter cake increases. After a certain time of use the filter cake has to be removed from the filter, e.g. by backflushing. If this is not accomplished, the filtration is disrupted because the flow resistance of the filter cake gets too high, thus too little of the mixture to be filtered can pass the filter cake and the filter plugs. The specifications of the filter cake dictate the filtration method of choice

Fine Bubble Diffusers

Fine bubble diffusers are a pollution control technology used to aerate wastewater for sewage treatment. They produce a plethora of very small air bubbles which rise slowly from the floor of a wastewater treatment plant or sewage treatment plant aeration tank and provide substantial and efficient mass transfer of oxygen to the water. The oxygen, combined with the food source, sewage, allows the bacteria to produce enzymes which help break down the waste so that it can settle in the secondary clarifiers or be filtered by membranes. A fine bubble diffuser is commonly manufactured in various forms: tube, disc, plate, and dome.

Bubble Size

The subject of bubble size is important because the aeration system in a wastewater or sewage treatment plant consumes an average of 50 to 70 per cent of the energy of the entire plant. Increasing the oxygen transfer efficiency decreases the power the plant requires to provide the same quality of effluent water. Furthermore, fine bubble diffusers evenly spread out (often referred to as a 'grid arrangement') on the floor of a tank, provide the operator of the plant a great deal of operational flexibility. This can be used to create zones with high oxygen concentrations (oxic or aerobic), zones with minimal oxygen concentration (anaerobic) and zones with no oxygen (anoxic). This allows for more precise targeting and removal of specific contaminants.

The importance of achieving ever smaller bubble sizes has been a hotly debated subject in the industry as ultra fine bubbles (micrometer size) are generally perceived to rise too slowly and provide too little "pumpage" to provide adequate mixing of sewage in an aeration tank. On the other hand, the industry standard "fine bubble" with a typical discharge diameter of 2 mm is probably larger than it needs to be for many plants. Average bubble diameters of 0.9 mm are possible nowadays, using special polyurethane (PUR) or special recently developed EPDM membranes. Fine bubble diffusers have largely replaced coarse bubble diffusers and mechanical aerators in most of the developed world and in much of the developing world. The exception would be in secondary treatment phases, such as activated sludge processing tanks, where 85 per cent-90 per cent of any remaining solid materials (floating on the surface) are removed through settling or biological processes.

The biological process uses air to encourage bacterial growth that would consume many of these waste materials, such as phosphorus and nitrogen that are dissolved in the wastewater. The larger air release openings of a coarse bubble diffuser helps to facilitate a higher oxygen transfer rate and bacterial growth. One disadvantage of using fine bubble diffusers in activated sludge tanks is the tendency of floc (particle) clogging the small air release holes.

Groasis Waterboxx

Groasis Waterboxx is a plastic box designed to help grow trees in dry areas. It is developed by Dutch former flower exporter Pieter Hoff, and won the overall Popular Science Green Tech Best of What's New Innovation of the year award for 2010, ahead of the Apple iPad, Porsche 918 Spyder and many others.

Background

Large land areas in the world are too dry for trees to survive. Although water may be present in the ground, it is often too deep for small trees to develop a root structure to reach.

Design

The Groasis is a polypropylene bucket with a lid. It has a vertical tunnel in the middle for two plants. A wick allows water from inside the box to trickle into the ground. As a biomimicry, it is modelled after bird feces. As of 2010, the development has taken 7 years and cost $7.1 million.

Installation

A hole is dug in the ground by a human or a machine. Two plants are planted in the hole, and a cardboard is placed around the plants. A wick is inserted in the bottom of the Groasis which is then lowered over the plants and filled with water. Two lids are put on, funnels inserted and a cap plugs the top lid.

Function

It works as a plant incubator, sheltering both a newly planted sapling and the ground around it from the heat of the sun, while providing water for the plant. The lid collects water from rain and night-time condensation, which is then stored in the bucket. The water-filled Groasis releases small amounts (around 50ml per day) of water into the ground by a wick to water the tree and to encourage the tree to develop a root structure. The box acts as a shield for the water in the upper ground, and this water then spreads down and out instead of being drawn to the surface and evaporated. Both temperature and humidity beneath and inside the box are more stable night and day than without.

Applications

The box has been tested for 3 years at the Oudja University in Morocco where nearly 90 per cent of plants survived with the box compared to 10 per cent without. Apart from projects in warm arid areas, the box is being tested in wineries and cold mountain regions.

3

Managing of Wastewater Treatment Plants

Wastewater carries resources which have reached a dead end when discharged at the end of a pipe. Nutrients, for example, that were constituents of the food we ate only a few hours earlier are polluting our waterways instead of finding their way back into the food production process. Present conventional wastewater treatment systems are designed to protect the environment from eutrophication as well as from pathogens, however, only newer treatment plants are capable of recycling nutrients from the wastewater stream.

One such technology - 'Wastewater-fed Aquaculture' - was developed in Sweden by Dr. Björn Guterstam in the 1980s-. It is commonly thought that this ecologically engineered wastewater-fed aquaculture is better suited for use in a sustainable economy when compared to a typical conventional sewage treatment plant. To see whether or not this assumption holds true, wastewater-fed aquaculture (aq're) was compared with two conventional wastewater treatment plants (wwtp's). To facilitate such a comparison, the two wwtp's were selected within Trosa municipality in Sweden, south of Stockholm, where the aq're is also located. All three facilities discharge into the Baltic Sea and receive domestic wastewater of similar composition.

The comparison is based on calculation of so-called 'Sustainable Process Indices' (SPIs), highly aggregating measures for sustainable process evaluation with area (square metres) as their basic unit for comparison (Krotscheck 1995). A more detailed explanation of the adopted SPI methodology used for this study follows. For the aq're itself, three SPIs were calculated: One each for its first and second periods of operation and one for a third, theoretical (future, optimised) period of operation. This was done to reflect the trend of steady improvements being achieved in Stensund through permanent research efforts in this experimental facility. One SPI was calculated for each of the conventional treatment plants at 1997 conditions only, because it was assumed that there is little room for improvement for those rather well established conventional technologies compared to the aq're. Besides, the conventional wwtps have already been modernised twice in the past.

Physical Specifications

An ecologically engineered demonstration plant for wastewater treatment was constructed and operated in the seaside municipality of Trosa, Sweden from 1989 to 1999. To be precise, the plant was a wastewater-fed aquaculture inside a greenhouse with the purpose of recycling resources contained in the wastewater of a college, especially for nitrogen, phosphorus and heat energy.

The aquaculture operated in a Nordic climate (60°N), exposing it to extremes in both temperature and light regimes between summer and winter. Only its special architectural design (by Bengt Warne) and location on a southeast facing slope on the coast of the Baltic Sea allowed it to operate throughout the year, by making the best use of incoming solar radiation. Onc hundred and eighty square metres of three-layer glass, together with greenhouse curtains made of aluminium foil for shading during the summer and for insulation at night and during dark winter days, helped to create and maintain a year-round indoor subtropical climate ideal for organisms to thrive. Eight treatment steps treated the water in the aq're to recreational standards. After primary settling, the wastewater was collected in a storage tank. From this tank, the water was pumped into the greenhouse. In an anaerobic tank, organic compounds were degraded and metal sulfides precipitated by sulfur-metabolizing bacteria. An aerated biofilter continued the microbiological mineralisation and detoxification of the wastewater.

In the next step, phytoplankton was cultivated (Ankislrodesmus, Scenedesmus and Chlorella). Primary production was the first step in the aquaculture's constructed food chain. Cultivation of zooplankton followed next (Daphnia, Copepoda, Rotifera, Ostracoda, Protozoa). These zooplankton functioned as grazers and detritus feeders, consuming the organic material from the previous steps. Floating plants were also grown on the top of this tank. Farming of fish (Tilapa marie, Sarotherodon sp., Colossoma sp., Pterygloplichthys gibbiceps, Cyprinus carpio, Carassius auratus), tropical aquatic plants (Eichhornia crassipes, Pistia stratoides, Hydrodeis nymphoidestakes), ferns (Azola filiculoides, Salvinia auriculata), and the temperate duckweed (Lemna minor) took place in the next tank. This type of fish policulture was inspired by Chinese aquaculture practices. Hydroponic cultivation of plants, especially tomatoes (Lycopersicon sp.) took place in five parallel 'hydroponic channels'. In a last step, a water cascade built out of flow forms designed to aerate the water, led the treated water into an outdoor pond that discharged into an energy forest (Salix dasydados) located right next to the Baltic sea. The aquaculture's plant production capacity on 80 m^2 of indoor growth area was directly related to the light availability and intensity in the aquaculture. Artificial lighting needed for plant production during the darkest time of the year consumed up to 11.5 kW over 15 h of operation per day. Plants were harvested for sale, fed to fish and pigs and surplus was composted.

Because the energy budget of the aquaculture was important, heat-energy was captured using heat pumps (3.3 kW electrical equivalent and 9.5 kW heat equivalent). The heat pumps recaptured heat from both indoor air and water throughout most of the year. The 160 m^3 of water contained within the greenhouse acted as an important heat buffer, and were essential for the heat-energy budget of the aq is. Despite the burning of 2001 of oil per year to supplement the heating of the aquaculture during its first period of operation, a portion of the recaptured heat energy was exported to the college. During its first period of operation, one person worked full time (40 h per week) in the aquaculture. During the second period of operation, one person worked 20 h per week. For the optimised third period of operation, one person was assumed to also work 20 h per week.

Lagnö and Trosa Conventional Wastewater Treatment Plants

Lagnö (constructed in 1977 and modernised in 1991) and Trosa (constructed in 1963 and modernised in 1987 and 1992) are two conventional wwtps also located within the Trosa municipality. Langnö serves 250 person equivalents (p.e.) and occupies 135 m^2 of land area, whereas Trosa treats the sewage of 5300 p.e. and occupies 590 m^2. Both wwtps use conventional technologies; a mechanical treatment step for sludge separation, a biological treatment step for COD/BOD reduction, and a tertiary step using chemicals for phosphorus precipitation. Trosa uses 85.5 t of iron chloride per year and Lagnö uses 5.41 of poli-aluminium chloride per year to precipitate phosphorus. The Trosa plant also burned 4 m^3 of oil per year for heating purposes, whereas the Langnö plant did not require any oil for heating. Only the Trosa plant operated a heat pump. With it, 788 000 kWh of heat energy were generated in 1996, of which about half was exported to the city. At Trosa one person worked halftime (20 h per week), while at Langnö one person only worked for 8 h per week.

The Sustainable Economy

The concept of sustainable development has been widely discussed since its introduction to the public in the Brundtland Report. Sustainability is a long-term concept for our living environments to guarantee our present quality of life and the accessibility of natural resources, without allowing long-term ecological damage that would compromise ecosystem integrity and the quality of life of future generations. In a sustainable economy, to realise the concept of sustainable development, the following system conditions should be considered:

- Renewable resources can only be used at a maximum rate that does not exceed their rate of replenishment.
- Anthropogenic material flows cannot exceed the local assimilation capacity of the biosphere.
- Non-renewable resources can only be used at a rate needed to find sustainable substitutes.
- The natural variety of species and ecosystems must be sustained or improved in order to maintain or increase system stability.

These conditions encourage material and energy efficiency and recycling. They also imply the fact that solar energy will be the only available energy source for a sustainable economy in the long run. The utilisation of solar energy is bound to surface area. Area, in turn, is also a limited resource because our planet is finite. Hence, in a sustainable economy, area and solar energy would be the limiting factors to growth within a given region. This is comparable to a forest, where biomass production is limited by the availability of light and area for photosynthesis.

The Sustainable Process Index (SPI)

The SPI is a measure developed to evaluate the viability of processes and services in a sustainable economy by aggregating their life-cycle environmental burden into the limited resource 'area'. Calculations for the SPI can help to identify areas required to provide the raw materials and energy of a process or service and the areas required to accommodate their by-product flows in a sustainable manner. Area in m^2 is the basic unit in all SPI calculations. Again, the assumption is that in a sustainable economy the only real income of our planet is the fairly constant amount of solar energy and that the conversion of solar energy into goods and services is bound to area.

Relating the total area needed for a process to the statistical average area available to one person in a region, results in the fraction of the area available to one person which is needed to sustainably embed this process into the region. This fraction also tells how much more area is left to sustainably embed all other processes this person needs. In a sustainable economy, one person cannot use more area than what is available to him or her. The sum of all fractions of all processes needed by one person must not be larger than one. It will only be then that he or she lives a truly sustainable life.

Specifically, in this study, the SPI expresses the fraction of land needed to sustainably dissipate a person's yearly load of sewage in the Trosa municipality depending on the treatment technology in use. The statistical average area available to one person in the Trosa municipality is 20 000 m^2

per capita. The SPI becomes a measure of the expense of sewage discharge depending on the treatment system. Excel, Version 5.0 was used for all calculations.

The Indoor Wastewater-fed Aquaculture

First Period

In its first period of operation, during 1990-1993, the aquaculture, just like most research facilities, had some problems in getting started. Not enough was known about how exactly the parts of the system should be dimensioned and arranged, and how the plant should be managed. Only one heat pump was installed during the first operation period. 2001 of oil were needed for supplemental heating per year during the winter.

Outflow was low, on average 6 m^3 of treated water per day. Reduction of total nitrogen did not exceed 60%, and total phosphorus reduction was only 45%. 1000kg of chemical phosphorus precipitant was used (poli-aluminium chloride). Thus, during the first period of operation, the SPI did turn out less favourably for the aquaculture. Calculations show that one person needed approximately 16% of his or her available living area in this period to sustainably dissipate his or her annual amount of wastewater into the Baltic when relying on the wastewater-fed aquaculture's services.

Second Period

Ambitious experiments soon made progress possible. During the second period of operation (1994-1997), nitrogen was reduced by about 69% and phosphorus by about 72%. Two heat pumps were operated to recycle heat energy in this period. No oil was needed for additional heating. However, still the same amount of chemical precipitation agent was needed as in the first operating period because natural precipitation processes could still not get started. Outflow of treated water in this period averaged 12 m^3 per day. One person needed approximately 6% of his or her available living area in this period to sustainably dissipate his or her annual amount of wastewater into the Baltic when relying on the wastewater-fed aquaculture's services.

Third Period

During the third period of operation, a theoretical scenario of operation with an optimised treatment system, nitrogen reduction is assumed to be as high as 90% and phosphorus reduction 95% with natural precipitation processes functioning. Four heat pumps will be operating. A higher daily output of 20 m^3 of treated wastewater is thought to be realistic. Considering these improvements, one person needed approximately 3% of his or her available living area in this period to sustainably dissipate his or her annual amount of wastewater into the Baltic when relying on the wastewater-fed aquaculture's services.

Langno and Trosa Conventional Wastewater Treatment Plants

Langnö

The smaller of the two conventional treatment plants has an average daily outflow of 50 m^sup 3^ of treated effluent per day. Approximately 40% of the total N and 95% of the total P contained in its sewage stream are recovered. 5400kg of poli-aluminium chloride are used annually as a chemical precipitation agent. Langno does not have any heat pumps installed. One person needs approximately 29% of his or her available living area to sustainably dissipate his or her annual amount of wastewater into the Baltic when relying on the Langnö sewage treatment plant's services.

Trosa

The larger of the two conventional treatment plants has an average daily outflow of 1600 m^3 of treated effluent per day. Approximately 40% of the total N and 97% of the total P contained in its sewage stream are recovered. 85 500 kg of iron chloride are used annually as a chemical precipitation agent. Heat pumps are in place to recover heat energy. One person needs approximately 36% of his or her available living area to sustainably dissipate his or her annual amount of wastewater into the Baltic when relying on the Trosa sewage treatment plant's services. The results of this study indicate that the treatment of wastewater with any of the evaluated facilities is more sustainable than not treating wastewater at all. It could be shown, however, that the ecologically engineered indoor wastewater-fed aquaculture could compete with the conventionally engineered treatment facilities in all of its operating periods. The main reasons for such a good SPI for the aquaculture were the high rate of nutrients recovered and its multi-functionality. The use of heat pumps and reduced use of chemical precipitants in the aquaculture also played a positive, but less significant role.

Concluding, the indoor wastewater-fed aquaculture is favoured by the SPI concept over the two conventional treatment technologies discussed in this chapter. This is especially astonishing, considering the fact that the aquaculture is only a prototype, whereas the Lagnö and Trosa wwtps are well established conventional technologies with presumably less potential for process optimisation. Future wastewater management systems designed and operated under ecologically sustainable criteria such as aquaculture are foreseen to decrease the total SPI even more as they will recover more nutrients, reclaim more energy, and supporta multitude of other uses. Besides sewage treatment, they will provide greenhouse space to produce ornamental plants and vegetables, aquarium space to produce fish and algae, and space for guided tours and lectures. They will not impact the environment, instead they will replace parts of today's consumed amounts of commercial fertilizers and fossil fuels, and teach visitors about the principles of sustainability.

Three Sampling Methods for Aerosolized Endotoxin

Endotoxin is a structural component found in the outer cell wall of Gram-negative bacteria. Various human cells have been found to respond to endotoxin. It appears, however, that the most important physiologic reactions are those that occur in cells of the immune system. Some of the biologic responses associated with cellular reactions to endotoxin include fever induction, macrophage activation, B-cell mitogenicity, restrictive lung function changes, and reduction in alveolocapillary gas transfer. Previous research studies have documented human exposure to endotoxin-containing dusts and aerosols. Several occupational studies have associated the presence of adverse health effects among agricultural workers exposed to organic dusts laden with endotoxin.

Other studies have suggested that endotoxin-contaminated water aerosols, such as those generated during sewage treatment or atmospheric humidification, may be the source of upper respiratory and gastrointestinal disorders observed in an exposed population. Accurate quantification of airborne endotoxin is necessary for correlation of endotoxin levels with observed health effects.

In addition, a standardised endotoxin sampling method is needed to appropriately establish acceptable versus unacceptable exposure levels. The results of past research have shown this to be a difficult task.

Several studies have demonstrated that the amount of endotoxin detected when different sampling methods are employed is highly variable and can be affected by factors such as sampling media (filter types), analytical methods, and aerosol composition.

Further, Milton and co-authors and Thorne and co-authors have suggested that endotoxin binds with varying degrees to different filter materials, enhancing the variability in endotoxin recovery during analytical extraction. Given these findings, this study was undertaken to provide data that will help in establishing a standardised sampling and analytical method that can be used to accurately estimate human exposure to aerosolized endotoxin.

The purpose of this study is to compare three different sampling methods for the collection of aerosolized endotoxin. Different collection media are employed in each method and consist of a liquid impinger, a glass fibre (GF) filter, and a polycarbonate (PC) filter. The GF and PC filter sampling methods were selected as comparative media since they have been shown to give consistent results in sampling for endotoxin-containing aerosols.

The liquid impinger sampling method was selected because it does not involve an extraction step during laboratory analysis, which allows the method to be used as a control against the variability seen during analysis when filter collection media are used in endotoxin recovery.

Materials and Methods

Endotoxin samples were collected outdoors, near the base of a 30-inch-high spillway in the primary clarifier of a sewage water treatment facility. This sampling site was selected because it is an environment in which endotoxin concentrations are likely to be relatively high. The impinger sampling train consisted of an SKC[R] 25-millilitre (25-mL) midget impinger with a standard nozzle, attached to a Gilian[R] GilAir[R] personal sampling pump calibrated at a flow rate of 1.60-1.64 litres per minute (L/min). For the collection of endotoxin, each impinger contained 20 mL of pyrogen-free water. Prior to sampling, all impingers were dry-heat sterilized for one hour at 250[degrees]C. The sampling trains for the filter sampling methods consisted of either a GF filter (1.0-micron [1.0-[micro]m] pore size, binder-free) or a PC filter (0.4-[micro]m pore size) housed in an SKC 37-millimetre (37-mm) polystyrene filter cassette holder, attached to a Gilian GilAir personal sampling pump calibrated at a flow rate of 2.06-2.10 L/min.

Because of the potential for contamination from atomised water in the proximity of the spillway, all filter sampling was performed closed-faced. Side-by-side endotoxin sampling was performed by positioning a liquid impinger, a PC filter, and a GF filter sampling train (as a triplicate set) 3 inches apart with their inlets 15 inches above the bottom of the spillway, 12 inches from the wall of the falling water, and at 45 degrees relative to the surface of the water. Two sampling runs were performed, each 30 minutes in duration and each consisting of five triplicate sets that collected endotoxin at different locations along the base of the primary clarifier spillway. The two sampling runs resulted in the acquisition of 10 samples for each endotoxin collection method. A time period of three hours elapsed between the two sampling runs. At the time of the sampling, the outdoor air temperature was approximately 24[degrees]C and the relative humidity was approximately 35 percent.

The authors extracted endotoxin from the GF and PC filters by placing each filter in a vial containing 5 mL of sterile, pyrogen-free water (at 37-40[degrees]C) for 60 minutes and sonicating each vial at 10-minute intervals. The endotoxin in all samples was quantified with the Kinetic Chromogenic Limulus Amebocyte Lysate (LAL) Assay (Nelson Laboratories, Salt Lake City, Utah). This assay places 0.1 mL of either the liquid impinger solution or the filter extract in a microplate along with 0.1 mL of LAL reagent. The mixture is incubated for one hour in a spectrophotometre at 37[degrees]C. Spectrophotometre results are compared to control results using standard endotoxin dilutions, and detected endotoxin levels are reported in endotoxin units (EUs).

In an effort to determine if interfering materials were present on the filter media or in the impinger solution, tests for enhancement or inhibition were performed as a part of the LAL analytical procedure. The results of these tests showed that the percentage recovery of endotoxin from spiked samples was normal (between 50 and 200 percent), suggesting that no enhancement or inhibition of biological activity had taken place.

Blank filters and impinger solutions were analysed for contamination with endotoxin during each assay and the amount of endotoxin on the blank media was subtracted from associated samples.

The results show that for both runs the mean amount of endotoxin detected in the impinger samples was two to three times greater than the amounts obtained with GF filters and approximately 10 times greater than the amounts from samples obtained with PC filters. In an effort to determine the degree of variation for a given sampling method during each run, the method's standard deviation was used to estimate its coefficient of variation (CV).

For Run 1, the GF filter method provided a lower CV than the PC filter and impinger methods (0.29, compared with 0.99 and 1.46, respectively). For Run 2, the difference between the CVs was not as pronounced. In Run 2 the impinger method had the lowest CV (0.47), compared with that for the GF and PC filter methods, which had similar CVs (0.68 and 0.72, respectively).

In an effort to determine the presence of a correlation among the results of the three methods, a Spearman rank order correlation test was performed on the data for Run 1 and Run 2 combined. The results of this statistical test are provided and show an apparent correlation between the sampling results obtained with the impinger method and those obtained with the GF filter method. This non-parametric analysis was performed instead of a Pearson correlation coefficient test because of the small sample size used in this study.

The results show a large decrease between Run 1 and Run 2 in the levels of endotoxin detected with each sampling method. One possible explanation for this observation is that a rainstorm produced approximately 3/4 inch of rain during the three hours that elapsed between the runs. It may be that the decrease in the collection of airborne endotoxin was caused by a dilution of the wastewater in the primary clarifier, or it is possible that the rain physically removed endotoxin-contaminated aerosols from the air.

The data also show that the highest endotoxin levels were obtained with the impinger sampling method. This result was expected because the method does not involve an extraction step during laboratory analysis, most likely enhancing the recovery of endotoxin during the analytical procedure.

It should be noted, however, that the use of impingers does have limitations. Impinger samplers are often considered inconvenient for use as field-sampling devices. If the impinger is tipped during sampling, the sampling solution inside the impinger may spill, resulting in inaccurate results, or the solution may become contaminated by physical contact with the outside environment. An additional source of contamination can be the required transfer of impinger solutions before and after sampling.

In addition to facility of use in the field, another important aspect in the choice of an endotoxin sampling device is good precision, or high reproducibility of results. Most studies comparing endotoxin aerosol sampling methods have not evaluated the precision of different methods.

The CVs estimated in this study provide a comparative measure of precision for each sampling method. Evaluation of the CVs suggests that when sampling takes place during conditions of relatively high levels of endotoxin (Run 1), the GF sampling method provides better precision. In contrast, while sampling is conducted during conditions of relatively low endotoxin levels (Run 2), a smaller difference among the CVs for each method was observed.

Because of the small sample size in this study, the statistically significant correlation found between impinger and GF filter results does not conclusively suggest that endotoxin levels obtained with a glass fibre filter can be directly correlated to those obtained with an impinger. This result does, however, provide the basis for further investigation, which could clarify the appropriateness of applying a correction factor to the results obtained by use of either method, especially when these methods are used to associate sampling results with human health effects.

As with previous research performed to compare the results of different sampling methods for the collection of endotoxin-laden aerosols, the study reported here detected different endotoxin levels when different sampling methods were used. The results show that endotoxin levels were consistently highest with the impinger sampling method, corroborating the theory that the recovery of endotoxin from sampling media is enhanced if extraction is not an element of the analytical procedure.

Nevertheless, the fact that the impinger sampling method carries an increased risk of sampling error when used in a field setting, combined with the reproducible results produced by the GF filter sampling method in Run 1, means that this research lends some support to the conclusion by Wood and Jacobs (1997) that GF filters should be designated as the standard method for sampling endotoxin-contaminated aqueous aerosols. Another important finding of the study reported here is the apparent correlation between the results obtained with the GF and impinger sampling methods. While the limited sample size does not allow for direct conclusions to be drawn from this observation, the finding does validate the need for larger studies to more accurately identify the true variability among these sampling methods and to determine the appropriateness of employing a correction factor when the results are compared.

Bioaerosol Dispersion from Wastewater Treatment

The Federal Water Pollution Control Act Amendments of 1972, known as the Clean Water Act, set a national goal of eliminating pollutant discharges to navigable waters by 1985. Some communities, industries, and private entities have met this goal by recycling treated water to irrigate agricultural crops or open spaces. While this approach eliminates direct discharges to surface water, sewage treatment plants and spray irrigation are a potential source of bioaerosols that may disperse downwind. This identifies appropriate management practices that reduce the risk to human health posed by exposure to bioaerosols from treatment and irrigation of wastewater. A recent literature review provides a comprehensive examination of research on the generation of bioaerosols from the treatment and land application of wastewaters and biosolids. Bioaerosols include bacteria, fungi, viruses, and identifiable cellular remnants. In all cases, bioaerosols require a carrier for air dispersal, such as water, soil, or other particulate matter.

For the study reported here, key research reports cited in the literature review by Brooks and co-authors were examined to identify both the characteristics of the treated wastewater and the wastewater management practices that led to detected bioaerosol dispersion. This review reveals measures that may be taken to limit the generation and dispersion of bioaerosols.

Teltsch and Katzenelson studied a single sprinkler that sprayed wastewater treated with a combination of primary sedimentation followed by trickling filter, sand filtration, and chlorination in sequence (1978). The reported $BO[D_5]$ level was 30 milligrams per litre (mg/L). The concentration of coliforms in the effluent was $[10^3]$ to $[10^5]$/mL. To test for bioaerosol dispersion, Teltsch and Katzenelson introduced marker bacteria (a mutant E. coli) into the wastewater at a similar concentration ($[10^5]$/mL).

The reporting unit for fecal-coliform testing for regulatory compliance is the number of colony-forming units (CFUs) per 100 mL of water. If all of the mutant bacteria in this experiment were viable, the effluent concentration would be $[10^7]$ CFUs/100 mL. Common fecal-coliform standards for disinfection of wastewater used for spray irrigation in the United States range from 2.2 to 200 CFUs per 100 mL. The bacteria levels in spray effluent examined by Teltsch and Katzenelson are five to six orders of magnitude greater than the limits set by these standards.

Air samples collected 20 metres downwind of the sprinkler showed concentrations of viable bacteria in the air averaging 11 CFUs per cubic metre of air. Teltsch and Katzenelson note that the relatively short sample times (15 minutes for air samples) might have reduced the detectability of bacteria when they were at low concentrations in the effluent. Sampling for viruses in the air showed echovirus to be present in 4 of 12 samples. That research established that low humidity and high solar radiation reduce bacteria survival in bioaerosols. Bacteria concentrations in air increased 10-fold during night-time conditions when wind speeds and temperatures were lower, humidity was higher, and no solar radiation was present.

Teltsch and Katzenelson concluded that as the bacterial concentration in the wastewater increased, the probability of discovering bacteria in the air increased proportionally. The inverse was also true; reducing bacterial concentrations in wastewater reduced the probability of discovering bacteria in the air. They noted that bacteria were found in the air during daytime only when the concentration in wastewater was [10^3]/mL or greater. No bioaerosols were detected 20 metres (66 feet) downwind when marker bacteria diminished to 3.4 to 5.7 X [10^3] bacteria/mL or less (1978).

In a later report, die-away kinetics ([lambda]) of aerosolized bacteria were measured. In this study, the bacteria tracer was tritium, the irrigation apparatus was a jet fog sprayer with 100 percent aerosolization, and air samples were collected 20 to 40 metres downwind of the jet fogger. Complete aerosolization and a single sprayer would be expected to increase detection of downwind concentrations of microorganisms.

The report cautioned that values of lambda could not be used to predict downwind concentrations of microorganisms in an effluent spray irrigation field. Nevertheless, useful principles in die-away kinetics were discovered.

First, in conditions of low humidity and high solar radiation, a rapid die-away of bacteria occurred in the first 10 to 20 seconds after release. There was a 90 percent reduction for every 10 seconds of travel time in the afternoon experiments. At a wind speed of 20 miles per hour, completely aerosolized water would move about 30 feet per second.

Therefore, a 90 percent reduction in bacteria concentration could be predicted (ignoring Teltsch's caution) in 10 seconds, or at a travel distance of 300 feet, during "afternoon" conditions. In the 1980 report, Teltsch and co-authors also confirmed that die-away was much slower in conditions of high humidity when essentially no solar radiation was present (Teltsch, Shuval, & Tadmor). In subsequent research, Teltsch, Kedmi, and co-authors (1980) compared the survival ratios of enterovirus, coliform, and Salmonella found in air downwind of a wastewater spray system using three different sizes of sprayers.

Total-coliform levels in wastewater sprayed by the sprinklers ranged from [10^6] to [10^7] CFUs/100 mL. These concentrations are within the range of values reported for raw wastewater. Salmonella concentrations in the wastewater are reported as the most probable number (MPN) per 100 mL, and they ranged from 0 to 60, with an average of 10, which is also within the range for raw wastewater. Total-enteroviruses samples were reported as plaque forming units (PFUs) per litre and ranged from 0.0 to 8.2 X [10^4] PFUs/litre, with an average of 6.9 X [10^3] PFUs/litre. Samples were collected from 40 to 200 metres downwind of the various sprinklers used at the test site. Teltsch and co-authors concluded that enteroviruses and Salmonella survived downwind dispersion more successfully than total coliform.

Camann and co-authors prepared a comprehensive assessment of a Lubbock, Texas, land application system. Twenty-two centre pivot irrigation rigs were installed and operating at the 3,665-acre Hancock Farm near Wilson, Texas, 15 miles from the source of the wastewater in Lubbock. The centre pivot rigs vary in length (i.e., radius) from 307 to 476 metres (1,007 to 1,562 feet). Operations commenced in 1982. An average daily flow of 3.7 million gallons of trickling filter effluent was delivered directly to the irrigation rigs or to three storage reservoirs at the farm site. Varieties of microorganisms were analysed. Both upwind and downwind areas were measured to provide a comparison with background conditions.

Composite wastewater samples were collected at the individual centre pivots before each air-sampling exercise was conducted. Air samples were collected 10 metres upwind of collection sites, and downwind samples were collected from 50 to 400 metres from the spray apparatus. Different types of sampling apparatus were used simultaneously during the research, and various protocols were established for quality control. The report noted spray irrigation aerosol rates of 0.26 percent to 1.04 percent according to dye test samples from 25 to 80 metres downwind.

Total-organic-carbon concentrations of 105 mg/L and total-suspended-solids concentrations of 149 mg/L characterised wastewater delivered to the reservoirs. Fecal coliform levels exceeded [10^6] CFUs/100 mL, and virus levels ranged from 100 to 1,000 PFU/litre. No data were provided on specific concentrations of the majority of microorganisms in pipeline water. Storage varied from as short as 60 days to as long as 180 days based on calculations taking into account flow patterns, rainfall, and other factors. The report notes that, reservoir impoundment significantly reduced the potential for exposure to microorganisms, citing reductions in coliform levels by as much as 99 percent and a lowering of viral levels to 10 PFUs/L. These observed reductions would presumably lower fecal coliform concentrations to [10^4] CFUs/100 mL, a value still well in excess of limits set by common regulatory standards for treated water used for spray irrigation (2.2 to 200 CFUs per 100 mL). This value also exceeds by an order of magnitude the effluent levels observed by Teltsch and Katzenelson that produce no detectable downwind concentrations of fecal coliform.

Camann and co-authors concluded that the effects of reservoir storage were significant and could be compared to the effects of a large buffer zone, estimated as 75 metres for fecal coliform and 300 metres for fecal streptococci, mycobacteria, coliphase, and C. perfringens. The latter three microorganisms are hardier than fecal coliform. Frequent detection of fecal streptococci in upwind samples indicates that this bacterium may be useful for comparison of downwind and upwind areas. Upwind air samples frequently contained fecal streptococci (27 of 30 samples).

The report includes precise values for fecal-streptococci concentrations in reservoir water, which are two orders of magnitude lower than those in pipeline wastewater. Interestingly, two other samples of pipeline wastewater chlorinated at the Lubbock treatment plant also contained fecal streptococci concentrations one to two orders of magnitude lower than those in pipeline water. None of the reservoir samples were subject to chlorination immediately prior to irrigation. The fecal-streptococci data provide an interesting prediction of the effects on detectable dispersion when wastewater has been disinfected more thoroughly. Three samples (M-20, M-16, and M-19) with fecal streptococci concentrations averaging 5 CFUs/mL (or 500 CFUs/100 mL) are of interest. Two came from reservoir water, and one came from pipeline wastewater chlorinated at the Lubbock plant. When upwind samples were analysed at the time this wastewater was sprayed, background concentrations of fecal streptococci averaged 0.2 CFU/[m^3]. Earlier samples collected to determine background microorganism densities in ambient air showed fecal streptococci concentrations averaging 0.47 CFU/[m^3]. Therefore, at this site, background concentrations of fecal streptococci may be expected in the range of 0.2 to 0.5 CFUs/[m^3].

For the three samples referenced (M-20, M-16, and M-19), the downwind concentrations of 15 of 16 fecal-streptococci samples fell within the range in the background samples. The outlying data point was a value of 0.7 CFUs/m at 40 to 59 metres downwind (131 to 169 feet). Data presented indicate that spray irrigation at the time these samples were collected often produced downwind concentrations of other microorganisms equal to or less than upwind concentrations. Three reports by Italian researchers address dispersion of bioaerosols from wastewater treatment facilities. Carducci and co-authors reported on a yearlong experiment that measured dispersion of microorganisms at an activated sludge plant that used turbine aeration (1999). In activated-sludge facilities, wastewater is introduced to turbine aeration tanks following primary settling. Active agitation of wastewater with surface aerators (turbine aeration) would likely produce bioaerosols from an open tank.

Sampling sites were established above the tank and at locations 20 metres and 50 metres downwind. The purpose of the experiment was to clarify relationships among viruses and bacteria at different distances from the tank. No readily discernible wastewater concentration data were reported; however, several findings are of interest.

First, the presence of reovirus and enterovirus at all sampling locations was a seasonal phenomenon. Measurable concentrations of these viruses were found only from June through October.

Second, viral presence consistently correlated only with fecal streptococci.

Third, over a distance of 50 metres, virus contamination dropped by 15 percent, while total bacteria count decreased by 88 percent, fecal streptococci by 98 percent, and coliphages by 99 percent.

In a second report, Carducci and co-authors reported on experiments at several wastewater treatment facilities in Livorno, Italy (2000). Preliminary testing identified 20 areas at the facilities with the highest likelihood of detectable microorganisms. The highest levels of contamination were found in a sewage-washing station along the collection lines, while the next highest concentrations were found near mechanical sewage-handling equipment, such as primary and secondary screens, and mechanical aeration of the grit chamber.

A sludge treatment plant had mid-range levels of bacterial contamination and high frequencies of samples testing positive for viruses. The areas around the aeration tanks did not appear to be highly contaminated, a finding that Carducci and co-authors ascribed to the use of subsurface aeration. Earlier studies had shown that surface aerators (the turbine aerators mentioned above) produced high levels of contamination. For viral detection, the Carducci and co-authors recommended the reoviruses because of their frequent presence. In a third report, also published in 2000, Brandi and co-authors reported on tests for microorganisms at facilities using different types of sludge oxygenation (2000).

The first facility used mechanical agitation of sludge, while the second used a fine-bubble diffused-air system. Dramatic differences were noted in microorganism dispersion downwind of the two types of tanks. The authors concluded that "aerobic digestion with a submerged microbubble system seems to pose little risk from airborne transmission of pathogenic bacteria and fungi to waste-treatment workers or local residents. For this reason, the conversion of mechanical oxygenation systems to submerged oxygenation systems of the sludge should be considered". The reports of Teltsch and Katzenelson (1978); Teltsch, Shuval, Tadmor (1980); and Teltsch, Kedmi, and co-authors (1980) suggest that treating and disinfecting wastewater to remove fecal coliform prior to spray irrigation is an effective means of reducing potential human exposure. Furthermore, the best timing and setting for irrigation of treated wastewater is during the day in low-humidity conditions. Testing of multiple microorganisms identified variability in die-away rates and suggests that total coliform is not an appropriate predictor of downwind dispersion of other microorganisms. Fecal coliform consistently died away at faster rates than other microorganisms.

Research by Camann and co-authors supports findings that fecal coliform are less hardy than other microorganisms. Camann and co-authors also noted that all microorganism concentrations decreased as distance downwind increased, and that the highest downwind concentrations were observed when irrigation occurred at night. The report also noted higher-than-average downwind concentrations when air speed exceeded 7 metres per second, or about 16 miles per hour.

Camann and co-authors (1988) also suggest significant benefits associated with wastewater storage reservoirs in providing natural disinfection (which is likely due to settling of solids and to oxygen transfer from the atmosphere to the top layer of the stored water). In addition, their research suggests that fecal streptococci may be a better indicator organism than fecal coliform because of higher survivability and incidence in background concentrations. The relatively low downwind concentrations in water sprayed from the reservoir (or from pipeline water that had been chlorinated) indicate a substantially lower exposure risk when wastewater is more thoroughly treated.

Carducci and co-authors (1999, 2000) and Brandi and co-authors (2000) have substantiated the value of fecal streptococci as an indicator organism, noting that they were consistently associated with detection of viruses. Their work also establishes that reovirus and enterovirus contamination may be a seasonal event.

Their research on bioaerosol dispersion at sewage treatment plants establishes the potential exposure risks of sludge management and the mechanical agitation of raw wastewater. Their observations regarding the reduced generation of bioaerosols when sludge is oxygenated with submerged aerators suggest that this effect is of such importance that they recommend consideration of conversion of mechanical aeration plants.

In all of these research settings, there was a poor level of disinfection in the treated wastewater. In many, fecal-coliform levels fell within the range of reported concentrations found in raw sewage. One of the cases presented (the three samples measured by Camann and co-authors) indicates that when water is disinfected more thoroughly the downwind concentrations differ little from background concentrations (1988).

Research on die-away rates indicates that viruses are long-lived, and downwind concentrations are not significantly diminished in the distances measured in the reports reviewed here. Clearly, the key to reducing viral dispersion is to reduce source concentrations to low levels. This review reveals several measures necessary to reduce the generation of bioaerosols at treatment facilities and at spray irrigation sites. For treatment facilities:

1. Design facilities to avoid the direct exposure of raw wastewater to the air.

2. Avoid mechanical agitation of raw wastewater unless in a controlled area within a buffered setting.
3. Employ subsurface aeration for oxygenation tanks, lagoons, or ditches.
4. Minimise sludge generation and handling.
5. Institute infection control at facilities that lack these design features—for example, respiratory protection, frequent washing of hands, daily showers, laundry service for uniforms, and disinfection of surfaces where human contact is likely, such as desks, counter space, and cabinets.
6. Locate facilities within buffer zones.
7. Treat wastewater reused for irrigation to minimise concentrations of a wide range of microorganisms, especially viruses, which appear to have the greatest longevity of all bioaerosols.
8. Consider use of storage reservoirs for natural disinfection prior to spray irrigation of treated wastewater.

For spray irrigation facilities:

1. Use wastewater that has been disinfected to minimise all microorganisms.
2. Establish buffer zones free of human habitation.
3. To the extent possible, avoid aerosolization—for example, by directing spray downward and using coarser spray nozzles at the periphery of centre pivot rigs.
4. Consider reservoir storage if it is not already provided.
5. Consider establishing background concentrations of fecal streptococci and conducting downwind monitoring to check for bioaerosol dispersion above background levels.
6. Avoid spraying in high wind conditions (> 20 miles per hour).
7. Spray in daytime conditions when possible.
8. Spray in low-humidity conditions when possible.

Wastewater Treatment Plant

The vice president of Guatemala's visit to the newly opened Jucanya wastewater treatment plant in early 2000 was meant as a very public display of government getting work done in the service of the people, part of a flurry of public works ceremonies staged to build support for the ruling party, Frente Republicano Guatemalteco. Whatever sense of cooperation and progress the scene conveyed, it masked layers of underlying conflicts and obstacles. For months before the plant's inauguration in November 1999, Panajachel's mayor had refused to officially receive the facility, which the municipal government would be expected to manage despite an absence of trained staff, operating manuals, and funds for maintenance. From his perspective, it entailed an added cost and responsibility the municipality was in no way prepared to assume. What explains this tension?

From a public health perspective, wastewater pollution in Panajachel had already reached alarming levels. Around half of Panajachel residents discharge their household water into the network of open ditches that cross the town, converting a network developed and maintained over generations for irrigation into an outlet for raw sewage. These highly polluted waters are still used to irrigate crops-presenting direct contamination risks-and they flow into the lake. Lake Atitlan is still considered the cleanest in Guatemala, in part because its unusual depth provides a sink to trap pollutants and because the area has relatively little industrial activity. Nevertheless, water quality on the shoreline is an increasing concern for public health professionals and for business leaders, who note that tourists are already being deterred by warnings in guidebooks concerning pollution in the lake.

Figure: *Water Treatment Plant*

Efforts to construct a modern wastewater treatment plant were initiated by former Panajachel mayor Dr. Sergio Lavarreda. A dentist by training, Lavarreda had grown up in the town and witnessed the transition from communal water sources to piped household water supply. As mayor, he made public health a priority and spoke of how untreated waste-water flowing into Lake Atitlan would quickly threaten the tourism industry. The European Union invited Lavarreda on a study trip to Europe, where he toured various facilities and decided a modern wastewater treatment plant was needed. In 1992, he requested assistance from the European Union to build a plant.

From an engineering perspective, the plant is well designed, incorporating features to minimise maintenance and even generate revenue through production of biogas and fertilizer. However, because plant operations were begun without the customary start-up phase, in which design adjustments are made in the field, it has several operating problems, including tank corrosion, pipe breakage, and loss of the biogas production capacity (Luis Villalobos, plant engineer, personal communication, March 2001; Arthur E. McGarity, engineer and professor, Swarthmore College, personal communication, September 2001). The plant is also operating well below its intended capacity because of delays in getting the town's residents and businesses to connect to the system.

Most worrisome to the municipal government is its inability to generate sufficient revenue through user fees to cover the costs of plant operation and maintenance. Monthly operations and maintenance costs in 2001 were 31,000 quetzales ($3,875 at Q8 per dollar), while fees actually collected by the municipality (some Q4,200 per month in the best case) covered only 13 percent of this amount.

If costs for plant replacement are factored in (current plant construction costs extended over 20 years, without inflation, total Q18,000 per month), user fees covered only 9 percent of the total operations and maintenance costs. Initial rules set by the municipality actually discouraged property owners from connecting to the system because they demanded that users demonstrate legal ownership of the property and be up to date with other tax bills.

These requirements were subsequently lightened: a user requesting sewage connection need only be current with payments for trash collection and water supply, then pay a connection fee of Q75 (about $9.40). Of some 1400 households recorded by the municipal government, slightly more than half had been approved for connection by April 2001, and 550 connections were operating. Three of the 8 major hotels were connected and paying their fees, as were 6 of some 40 guesthouses. Despite the clear public benefits of wastewater treatment, and despite a series of meetings convened to explain the new requirements, municipal authorities were facing a wall of noncompliance.

4

Challenges of Wastewater Management

In November 1999, with great fanfare, the Guatemalan government inaugurated a new wastewater treatment facility in Panajachel on the shore of Lake Atitlan, one of three principal tourist destinations in the country. Financed primarily by the European Union as part of its aid package for post-civil war reconstruction and highlighted by some in the national government as a symbol of social investment promised in the peace accords, the plant is state-of-the-art among public wastewater treatment facilities in Central America. Within months of its inauguration, however, it was threatened with closure.

The most immediate problem, common to aid-financed infrastructure projects around the world, is a lack of funds for operations and maintenance. Underlying this apparently straightforward financial obstacle is a challenge that runs much deeper, a failure to "build" the social and organisational foundations for project sustainability.

The Panajachel plant is designed to fulfill a "public need" as assessed by project planners, but for the first two years following its inauguration, there were few in the town who saw it as a public asset, much less as a commons-collectively owned, managed, and maintained. As recently as the 1950s, when Panajachel was still a rural, primarily agricultural community, the town's water infrastructure was managed as common property. With the extension of state authority and the gradual incorporation of most households into the market economy, fees paid to the municipality for public services replaced traditional household contributions of labour directly in service to the community to maintain the irrigation system. The current controversy surrounding the treatment plant points toward an unmet need to reconstitute community around a new common interest in meeting the relatively new problem of wastewater management. This chapter describes an ongoing program of action research aimed at generating public dialogue among a broad range of local groups and national agencies with a stake in the social, economic, and public health consequences of pollution in Lake Atitlan and in efforts to improve wastewater management and treatment.

In keeping with the action research model (Whyte 1984), the goal is not only to study a problem, but also to work collaboratively with local partners in assessing both the underlying causes of the problem and possibilities for cooperative action to address it. As such, the analysis presented here is a snapshot of a work in progress, incorporating information from interviews, group consultations, and discussions conducted periodically during January 2001-September 2002, including a stakeholder dialogue organised in April 2001 at the Atitlan Nature Reserve on the outskirts of Panajachel.

The action research process has been closely linked to the ongoing work of the Nature Reserve and its outreach with communities of the lake basin, and to an experiential education program of the Higher Education Consortium for Urban Affairs (HECUA). Guided by a philosophy of learning through reflective engagement in contemporary social problems, the study program was designed to include visits with many of the key players in Panajachel's wastewater management controversy, including the plant engineer, town mayor, head of the basin authority, a number of non-governmental organisation (NGO) leaders, and neighbourhood residents.

The program also hosted study visits by outside faculty, including an engineer who provided assistance to put the plant's laboratory facilities in operation. The authors conducted individual interviews with these and other stakeholders, drew on an existing body of prior interview data in reconstructing the historical context, and held discussions with key persons who had since departed the local scene, including the lead designer of the plant and the town's former mayor who championed the project. We compiled a briefing report on the wastewaler and water quality challenge, citing the best available data, and including a very basic analysis of the potential economic implications of the problem, the gap between user fees and operations and maintenance costs for the plant, and the equity of the fee structure.

The briefing report, in the absence of any official document on the issue available within the community, proved helpful to elicit feedback and triangulation of perspectives that improved our shared understanding of the problem-and to launch discussions on potential solutions. It specifically provided the background for the stakeholder dialogue forum, hosted by the Nature Reserve, which achieved agreement among many of the key figures on principles for addressing the problem, as detailed later under Reasserting Community. In addition to offering insights into the practical questions of how planning and design of public works projects can be made more viable through efforts to reassert and strengthen community, this research yields theoretical implications on the role of state and market in processes of disembedding and re-embedding community. While we agree with McCay and Jentoft (1998) that the institutionalisation of state administration and market integration can potentially undermine community institutions for managing common property, we do not see this outcome as either inevitable or irreversible.

The Panajachel case offers instances in which state and market forces have reduced the space for social interaction around the commons, eroding networks of community, and alternately, instances in which the market and state administration become foci for reasserting community and rebuilding institutions for managing the commons. The degree to which this potential is realised, we argue, depends upon the ability of social actors to engage in a dialogue that achieves sufficient alignment of shared interests among the multiple, overlapping subgroups whose relationships constitute the broader community.

Commons and Community: Situating the Argument

Our argument is grounded in a specific conception of commons and community, and of the dynamic and interdependent relationship between the two. To help distinguish this conception, it is useful to consider its place within the now extensive literature on common-property resource systems. Much of this writing takes as its impetus the influential views of Olson (1965) and especially Hardin (1968), who argued that individuals are incapable of managing collective resources for the common good without some externally imposed coercive threat. Consolidating a wide body of research providing evidence to the contrary, Ostrom (1990) presented the most systematic statement of how effective common-property resource systems have evolved and the mechanisms of "self-governance" that can sustain such systems. Each of these now classic texts is framed within the rational choice tradition, subscribing to the fundamental proposition that individuals act rationally to maximise their own utility in reference to institutional rules that provide incentives guiding individual choices. Ostrom's key proposition is that these rules often are designed and adapted quite effectively by the collective of users who depend on a shared resource, a possibility precluded by the imposition of privatization or state management-typical policy measures aimed at averting the "tragedy of the commons."

A more recent strain of research, while reaffirming the importance of common-property resource systems, takes exception with the limits of the rational choice tradition to explain either the social relationships that sustain these systems or the political and economic forces that threaten them. McCay and Jentoft (1998) provide a particularly succinct theoretical argument integrating both these strands of critique. Invoking Durkheim's concept of social solidarity, they argue that community is more than a collection of individuals or the aggregate of individual choices, that it provides an integrated moral framework: "the community can provide normative guidelines and meaning to the private sacrifices involved in collective action".

The vitality and persistence of common-property resource systems, which exert restraints on individual behaviour and incentives toward collective action for the common good, depend upon the broader network of relationships and social bonds that make up community. True tragedies of open-access resource use and ecological decline should be understood, they argue, as instances of "community failure": A working hypothesis is that the social conditions required for tragedies of the commons may result from situations where resource users find themselves without the social bonds that connect them to each other and to their communities and where responsibilities and tools for resource management are absent.

Amplifying concerns raised earlier by Bromley and Cernea (1989), McCay and Jentoft call for a focus on broader relationships of power that may undermine the social bonds of community, with particular attention to processes of state administration and market integration. The present study critically examines this link by asking how the assertion of state administration and rapid market integration has affected a long-standing common property irrigation system in Panajachel over the past two generations.

Yet, we also want to understand the roles the state and market play in the capacity of Panajachel's residents to address the relatively new problem of wastewater management. To undertake these analytical tasks, we need to move beyond the notion of a single community comprised of individuals with a common set of interests and subscribing to a single, integrated moral framework. However appealing such a view may be in the theoretical defence of traditional common-property systems, it fails miserably to portray the reality of Panajachel and, indeed, most places where common property resource systems are under stress today.

Community, in the sense we use the term, refers not to a place or to a fixed group of individuals, but instead to a set of shared interests, rooted in judgments of value, and to the networks of social relationships that form around these. These shared interests may comprise matters of collective identity, such as the observation of religious festivals in honour of a certain patron saint; they may also be matters of material welfare, such as the management of a forest or water resource on which community livelihoods depend. It is this collective valuing of something, and identifying a shared interest in its maintenance, that constitutes a "commons." Our use of the term "commons," therefore, includes, but is not limited to, the case of common-property resource institutions. The commons refers not to a physical entity as such, but to a "social event".

The creation and maintenance of a commons are expressions of community; maintaining the commons both depends upon the existence of social relationships and provides a focus for the continuation of those relationships. Following this conceptualisation, we must recognise that wherever the residents of a place or the stakeholders in a given project uphold a diversity of values and interests, we are dealing not with a single, unitary community, but with a mosaic of overlapping networks of community.

Panajachel and the Lake Atitlan Basin

The culturally diverse and ecologically rich region of Lake Atitlan in the Guatemalan highlands is in the midst of an economic transformation-one that is providing new opportunities to local communities while at the same time eroding the basis of many traditional livelihoods. In the past 30 years, following a regional trend, agricultural production in the lake basin has shifted from subsistence production of milpa (maize) and beans to intensive cash cropping using synthetic fertilizers and pesticides. In the last generation, cheap plastic tubes, used to carry water from springs and streams in the upper watershed for both household use and irrigation, and increased peasant household resources extended agricultural production into the dry season and into areas previously only devoted to seasonal planting. As a result, water use has increased and tensions are rising over its allocation and proper use. Tourism has also grown rapidly, particularly since the conclusion of the most violent period of the civil war in the 1980s and a return of relative security.

Hotels and small accommodations, restaurants, and arts and crafts stores are thriving, as are specialised bus and boat services. Tourism now provides wages for more than half of the economically active population of Lake Atitlan's northern shore towns and supplies the main market for most of the household arts and crafts produced by communities around the lake. Fifty years ago, apart from some merchants and the owners and operators of the three hotels on the lake shore, the vast majority of PanajachePs residents relied upon agriculture for their livelihood. Today, nearly 80 percent of economic activity in the town is derived from tourism. Tourism has also accelerated immigration to Panajachel and furthered a gradual diversification in its social composition. In 1964, approximately 3,000 people, most of whom were native K'aqchikel, lived in Panajachel.

By the time the most recent census was taken in 1994, the population had more than doubled. Today, Panajachel includes a large ladino population, along with K'aqchikel people from nearby towns such as San Jorge and Santa Catarina, Tzu'tujil people from across the lake, and K'iche and other Mayan peoples from other parts of the country, as well as some 700 foreign residents and an additional 3,000-4,000 tourists at peak season. In recent years, ecological decline has become more evident in the lake and the surrounding lake basin, whose population has grown to 257,000. Principal concerns are declining shoreline water quality caused by municipal sewage and agricultural runoff, deforestation and loss of shoreline habitat, and related threats to terrestrial and aquatic biodiversity.

Efforts to protect the lake basin are confronted with a classic problem of collective action: how can a set of actors with diverse interests be motivated to protect a common resource, especially when the burdens and the benefits of protection are not equally shared? Unlike past natural-resource management problems that could be effectively addressed at the village or town level, such as irrigation management or solid waste collection, reversing ecosystem decline and protecting water quality in the lake requires an additional level of cooperation at the basin scale. In view of this challenge, the Atitlan Nature Reserve has tried over the past decade to engage other local actors in providing a "platform to talk about our common future," a place where the various communities that live within and depend upon the lake basin can define issues of common interest.

The Nature Reserve, located just outside the town of Panajachel on lands of the historic Finca San Buenaventura, is a private concern providing opportunities for environmental education, recreation, and research. Its facilities are regularly used by nongovernmental organisations and government agencies dedicated to environment and development projects. The Nature Reserve has also sponsored meetings of the Mayan mayors of the lake basin to discuss issues related to solid waste management, biodiversity protection, organic agriculture, and other concerns. Current research on the problems of wastewater management in Panajachel is an extension of this engagement, made possible through existing links with neighbouring communities and with a variety of governmental and nongovernmental organisations within the lake basin. These efforts are described in more detail below.

Social History of the Jucanya Wastewater Treatment Plant

The vice president of Guatemala's visit to the newly opened Jucanya wastewater treatment plant in early 2000 was meant as a very public display of government getting work done in the service of the people, part of a flurry of public works ceremonies staged to build support for the ruling party, Frente Republicano Guatemalteco. Whatever sense of cooperation and progress the scene conveyed, it masked layers of underlying conflicts and obstacles. For months before the plant's inauguration in November 1999, Panajachel's mayor had refused to officially receive the facility, which the municipal government would be expected to manage despite an absence of trained staff, operating manuals, and funds for maintenance. From his perspective, it entailed an added cost and responsibility the municipality was in no way prepared to assume. What explains this tension?

From a public health perspective, wastewater pollution in Panajachel had already reached alarming levels. Around half of Panajachel residents discharge their household water into the network of open ditches that cross the town, converting a network developed and maintained over generations for irrigation into an outlet for raw sewage. These highly polluted waters are still used to irrigate crops-presenting direct contamination risks-and they flow into the lake. Lake Atitlan is still considered the cleanest in Guatemala, in part because its unusual depth provides a sink to trap pollutants and because the area has relatively little industrial activity. Nevertheless, water quality on the shoreline is an increasing concern for public health professionals and for business leaders, who note that tourists are already being deterred by warnings in guidebooks concerning pollution in the lake.

Efforts to construct a modern wastewater treatment plant were initiated by former Panajachel mayor Dr. Sergio Lavarreda. A dentist by training, Lavarreda had grown up in the town and witnessed the transition from communal water sources to piped household water supply. As mayor, he made public health a priority and spoke of how untreated waste-water flowing into Lake Atitlan would quickly threaten the tourism industry. The European Union invited Lavarreda on a study trip to Europe, where he toured various facilities and decided a modern wastewater treatment plant was needed. In 1992, he requested assistance from the European Union to build a plant.

From an engineering perspective, the plant is well designed, incorporating features to minimise maintenance and even generate revenue through production of biogas and fertilizer. However, because plant operations were begun without the customary start-up phase, in which design adjustments are made in the field, it has several operating problems, including tank corrosion, pipe breakage, and loss of the biogas production capacity (Luis Villalobos, plant engineer, personal communication, March 2001; Arthur E. McGarity, engineer and professor, Swarthmore College, personal communication, September 2001). The plant is also operating well below its intended capacity because of delays in getting the town's residents and businesses to connect to the system.

Most worrisome to the municipal government is its inability to generate sufficient revenue through user fees to cover the costs of plant operation and maintenance. Monthly operations and maintenance costs in 2001 were 31,000 quetzales ($3,875 at Q8 per dollar), while fees actually collected by the municipality (some Q4,200 per month in the best case) covered only 13 percent of this amount. If costs for plant replacement are factored in (current plant construction costs extended over 20 years, without inflation, total Q18,000 per month), user fees covered only 9 percent of the total operations and maintenance costs. Initial rules set by the municipality actually discouraged property owners from connecting to the system because they demanded that users demonstrate legal ownership of the property and be up to date with other tax bills. These requirements were subsequently lightened: a user requesting sewage connection need only be current with payments for trash collection and water supply, then pay a connection fee of Q75 (about $9.40). Of some 1400 households recorded by the municipal government, slightly more than half had been approved for connection by April 2001, and 550 connections were operating. Three of the 8 major hotels were connected and paying their fees, as were 6 of some 40 guesthouses. Despite the clear public benefits of wastewater treatment, and despite a series of meetings convened to explain the new requirements, municipal authorities were facing a wall of noncompliance.

The Social Context of Failure to Pay

Project planners commissioned by the European Union did not foresee problems in financing operations and maintenance of Panajachel's wastewater treatment system because they failed to understand the distribution of resources within the community. Nor did they grasp the historical and political obstacles to generating community support in public works and confidence in local government. Even if the tourist economy generates significant income for a community of this size, income inequality is huge across the spectrum from large hotel owners to the poorest residents. While the planners were correct in calculating that the burden of user fees would not be onerous on the basis of average income, they were unrealistic in assessing the burden on most households. Most residents of Panajachel have little or no income beyond what is required to meet basic needs.

More fundamental to the unwillingness of many of the town's residents to pay user fees was an enduring distrust of municipal government and its ability to manage funds in the public interest. This is due to generally low confidence in government institutions as a legacy of the civil war, as well as allegations of official corruption that plagued a series of recent local administrations.

Panajachel's mayor at the time the plant was inaugurated, Bracamonte, was impeached for efforts to "rescue" the lakefront from rich property owners, a ruling the Constitutional Court upheld, after which he sued to retain his position. The notion of an embattled public servant fighting to keep his title despite community opposition would have been unthinkable under traditional models of local governance. Traditionally, Mayan communities have modelled their relations with authorities on a practice of lifelong commitment, knowledge, and service to community. Under the current electoral system, however, candidates backed by the powerful national parties have typically had little or no experience of genuine service to the community and thus lack an ability to garner the legitimacy that comes from effectively representing community sentiment.

Waterworks and the Historical Transformation of Community

The maintenance of public waterworks, a point of such contention in Panajachel in recent years, was until not long ago a principal expression of community and collective responsibility. Panajachel's transformation over the last half century from a rural K'aqchikel agricultural community to an urban, multiethnic community dominated by the tourism industry was accompanied by a transformation in the relationship between community members and public works. The waterworks created to feed the highly productive tablon system of agriculture in Panajachel depended on seasonal service to community in which each household participated. While the irrigation ditches (tomas de tablones) that run between the raised tablones, or cultivation plots, were maintained by households, the large feeder canals that draw water from the rivers (tomas grandes), as well as the secondary canals that direct water to sections of tablones as required (tomas secundarias), were maintained collectively by the community in a festive day of work typically held on a Sunday at the beginning of the dry season (Mathewson 1984:95).

"There was one canal head (jefe de tomas) for the whole town," recounts Juan Ramos, a 71-year-old K'aqchikel elder. "He didn't get paid for his position (cargo). He had to be somebody who knew how to talk to people". The organising talents of the canal head were essential not only to plan the work itself but also to ensure full participation and resolve conflicts. Everyone who used the canals, he says, provided labour from their own household on these days of collective maintenance or paid for a worker in their stead. The shift to state administration of the municipal waterworks in Panajachel paralleled a gradual erosion in the traditional structures of alcaldia and cofradia, the interwoven civil and religious systems of local governance and service. Until 1975, the Mayan head of the alcaldia served as assistant (alcalde auxiliar) alongside the mayor, invariably a ladino, who holds formal authority in the eyes of the state. "Now there is no more native mayor (alcalde natural)" Ramos laments, "...today everything depends on a salary." Similarly, the expectation that households will provide service to the community (servicio a la comunidad) has been largely replaced with an expectation that they will provide a fee for service (pagar por el servicio) to the municipality. This includes a household reliance on municipal government to administer and maintain public waterworks. This shift of local governance authority from traditional community institutions to formal state institutions has been consistent in the lake basin and surrounding region over the past two generations (Seminario de Integration Social Guatemalteca 1968) and is linked to the economic and demographic transformations described above.

In Panajachel, as well as in some of the nearby villages, the traditional cofradia system that formalised and honoured service to community is dying out as many men in the next generation refuse to assume the positions of responsibility (cargos) this system entails. As more people move from an agricultural livelihood to the wage labour market, families also have less flexibility to provide their own labour as service.

The waning of collective work also entails a loss in social interaction, a form of "disembedding" that reflects both state and market influences. For communal waterworks, the annual day of maintenance not only enacted and reinforced community bonds as residents worked side by side on the canals and prepared food for the day. It also provided an immediate transparency in decision making over the project itself and consequently a sense of collective ownership.

Changes in the physical infrastructure of water supply have contributed to the changing character of community interaction. When piped water was first introduced by the municipality, it fed several wealthier households and businesses directly, as well as neighbourhood water reservoirs (pilas) used by most households. The pilas became a new meeting place for women who had drawn water and washed clothes together at the lakeshore. As more individual households gained separate piped water connections, these meeting places waned in significance. At the same time, water use rose quickly. It was no longer necessary to carefully conserve the water that previously was carried long distances. It also required payment of a monthly fee, adding to the growing list of goods and services (including electricity and, for many, food) that had to be purchased rather than produced within the household economy.

Reasserting Community

The problem of wastewater management in Panajachel has reached the crisis point over the last two years. Growing awareness has shifted the issue from the realm of everyday municipal management to high stakes politics. By no means resolved, the problem has elicited a range of responses from conflict to community activism and renewed collective action. Of particular concern here are two interrelated processes: the creation of new space for public dialogue and the reassertion of community networks to address the practical challenges of wastewater management.

We use the concept of "reasserting community" to indicate the creation and strengthening of social ties and interactions around common interests.

The social composition of the town has changed, new networks have become active, and new divisions have emerged. The nature of the resource management problem has changed as well: irrigation management involved the provision and allocation of water to a set group of users who derived a direct benefit. The benefits from participation in wastewater management efforts are mostly indirect and accrue to the town as a whole (though the economic and the health benefits are unevenly shared). At a broader scale, Panajachel's failure or success to manage its wastewater pollution problem represents one current in the larger challenge of protecting the ecosystem of the lake and, by extension, the livelihoods and welfare of the people who live around its shores. For each of these reasons, we are not simply suggesting that the old institutions be reinvoked, but that existing networks of community be engaged and new ones formed to meet the present challenges.

In retrospect, the task of building community awareness and dialogue clearly would have been much easier had it begun prior to the planning, design, and construction of the Jucanya wastewater treatment plant. When the finished plant was turned over to the municipality, the local administration, already faced with deep public skepticism and mistrust, saw it as an added burden. Underpressure from central government to make the project a success, the municipality first employed its conventional approach, by demanding payment from households and businesses for a service that few felt they needed. Faced with noncompliance, the municipality did organise a series of meetings, but, according to many we interviewed, these meetings focused on explaining a requirement as opposed to surfacing a problem and exploring potential solutions jointly. Many of the stakeholders who could have been working together to break the deadlock were more focused on assigning blame to others.

In this atmosphere of intense political maneuverings, we invited, with the approval of relevant authorities, a broad spectrum of stakeholders to the Atitlan Nature Reserve to assess obstacles as well as strategies for joint action. Prior consultations with a variety of local groups made it clear that those familiar with the wastewater problem were becoming increasingly concerned, and there was no functioning forum for these groups to explore how to work together. The premise of the meeting was that, while numerous shortcomings were evident in hindsight, the existing Jucanya plant represents an important asset for addressing the serious problems of pollution in Panajachel. The political tensions surrounding the treatment plant remained strong-neither the embattled Panajachel mayor at the time, Bracamonte, nor the former mayor who had initiated the project, Lavarreda (who had since become vice minister of the environment), came to the meeting. Apart from this, principal governmental agencies were represented, including the lake authority, Autoridad para el manejo sustainatable de la cuenca del lago de Atitlan y su entorno (Lake Atitlan Basin Sustainable Management Authority [AMSCLAE]);

official tourism agencies, Institute guatemalteco de turismo (Guatemalan Tourism Bureau [INGUAT]) and Camara de turismo (Chamber of Tourism [CAMTUR]), as well as nongovernmental actors, Amigos del Lago (Friends of the Lake) and Vivamos mejor (Let's Live Better); hotel owners; and other businesses represented by the Grupo gestor (Action/Management Association).

AMSCLAE is a seven-year-old interagency governmental body in which the municipalities of the lake basin, the governor of Solola, the military commander of the region, and the Chamber of Tourism are represented, along with a variety of NGOs working in the lake basin. Amigos del Lago and Vivamos mejor are both professionalised local NGOs. Vivamos mejor focuses on social services, such as health, education, and community development projects, and receives some external funding. Amigos del Lago, formed initially as a dues-collecting association to advocate the interests of the owners of upscale lakefront homes, is now helping to develop a master plan for the lake. Grupo gestor primarily represents major businesses in Panajachel, though its membership has expanded to include some small operations, such as those that dig sand from the riverbed to supply the construction industry.

For many of those gathered, it was the first opportunity to construct a comprehensive picture of the problem, with experts on hand to provide details of water quality conditions and health effects, the plant's operations, the extent of participation in the sewage system and the fees generated, as well as a detailed accounting of the actual costs for plant operations and maintenance. The format of the gathering, particularly the recounting of the history of public actions and accounting for the use of funds, and the opportunity for any participants to question and probe the presenters, paralleled the way communitywide meetings are still conducted in many of the K'aqchikel towns around the lake. The information shared in the forum made clear that the plant's operations were being heavily subsidised by the municipality and that even with full compliance, the current fee structure was inadequate to cover the plant's long-term maintenance. Participants also sketched out what they saw as plausible effects on the town's economy if failure to definitively address wastewater pollution should lead to a dramatic drop in tourism. Through these elements, the forum provided an incremental step in clarifying the domain of shared interests among the groups represented, as well as increasing the level of trust.

Participants agreed on two guiding principles: 1) polluters must pay, and since all residents contribute wastewater, all should pay; and 2) those who draw their income from tourism have a higher ability to pay and an obligation to protect the tourist destination-they must, therefore, contribute in proportion to their income. Meeting participants endorsed the following statement: It is clear that those who use the plant more and derive more economic benefit from its operation should pay a higher fee for the service than simple residential users. These differential fees should be based upon transparent criteria and should be made publicly known and developed jointly with those who will have to pay these fees.

Participants also agreed on a proposed alternative fee structure that differentiates among businesses according to the prices they charge their customers, increasing the burden on higher-end hotels and restaurants. Grupo gestor, while strongly endorsing measures to address water pollution, was at first hesitant about raising fees levied against the more expensive hotels. The group eventually did endorse the new fees, however, convinced that business interests were at stake. The group's main representative, the manager of the Hotel del Lago, in effect endorsed a fivefold increase in fees to his own business. If fulfilled under this schedule, user fees would cover over half the costs of operations and maintenance (disregarding future plant replacement costs), a fourfold increase from the amount of fees currently collected (though still requiring a substantial subsidy from the municipality). While by no means resolving equity concerns, the new proposed fee plan also reduces the equity gap. The three most exclusive hotels would together pay fees equal to nearly 1,200 household users.

The forum, along with documents that circulated before and afterward, generated a new commitment among many participants to build partnerships among government, civil society, and the private sector to address the challenge. As an example, Amigos del Lago, having learned of the specific conditions of the plant in preparation for the meeting, contacted a company in Canada about the ultraviolet bulbs that had gone unreplaced and explained the financial difficulties of the municipality. The company agreed to provide a one-time free set of replacement bulbs, valued at over $2,000. More important than such in-kind contributions over the long term is the role NGOs and business associations are beginning to play in generating broader public support for and participation in wastewater treatment efforts.

The role of the municipality also shifted markedly toward engaging this sort of community participation with the installation of a new mayor, Enio Urizar, in early 2002. Urizar, a schoolteacher, has ties to the grassroots civic committee that elected Bracamonte. In his own words, it works "to strengthen civil society" by organising to address problems at the neighbourhood level. Describing the election of his predecessor, he says "no one believed" that this grassroots movement would ultimately elect the municipal government in a race against organised national political parties. While embracing this commitment to the needs of the town's disadvantaged residents, he is also careful to distance himself from the confrontational approach adopted by his predecessor. By comparison, he says, "I am more politically attuned, in the sense that I work on the basis of consensus, with community participation."

This participatory, community-based approach has already proven remarkably successful in overcoming the deadlock of noncompliance in household connections to the wastewater treatment system. On the basis of an understanding gained from discussions with residents of the narrow alleyways and fringe neighbourhoods where exposure to open sewage is worst, Urizar and his team tackled the problem through collective rather than individual commitment. Many residents had voiced concern that the individual cost of connecting to the system was too high and that, in any case, the health benefits would only come to the neighbourhood when everyone was complying.

Under the new agreement, neighbours must form an alley committee in which all alley residents commit to connect to the sewage system. They must establish as well that all neighbours are up to date in payments for water supply and trash collection, along with the new sewage fee. The municipality provides plastic tubing to the residents, who are responsible for digging the channels and installing household and alley drainage lines that connect to the main system. As with traditional waterworks maintenance, those who do not provide their own labour may pay for someone to work on their behalf. As an added benefit, once the connections are completed, the municipality provides concrete tiles to pave the alleyways, which residents value as a significant improvement.

In March 2002, three densely populated alleyways were working with the municipality under this type of agreement to connect 185 additional families to the system. Community members felt they were investing in their own well-being, improving their neighbourhood, and helping solve the larger problem of pollution in the town as well. Many who supported this work in Panajachel claim, moreover, that where committees have formed to meet this shared problem, social networks have improved capacity to address other problems of collective action.

In parallel to this organising in residential neighbourhoods, business associations like the Association de comerciantes de Panajachel and the Grupo gestor are beginning to apply similar principles of collective responsibility and mutual accountability so business owners in specific sectors (hotels, restaurants, shops, etc.) will increase compliance and assume the costs of wastewater management. When viewed collectively, the economic rationale behind paying for a cleaner environment is overwhelming. As one active member of the hotel owners' group put it, "When we press for people and businesses to connect up to the wastewater treatment system, some say that we're being politicians, but we simply want tourism to continue in Panajachel.... We just want to do business".

This case provides an example of processes of community disembedding and rebuilding that call for more refined analyses than implied by a priori assumptions about the negative effects of state intervention or market integration. Community is not a fixed social unit, but a network of social interactions that are continually remade. When state administration reduces the space for social interaction around the commons, these networks erode. In Panajachel, local administration has taken over responsibility for municipal water provision and for maintaining the waterworks infrastructure. Community members no longer need to engage in relationships with neighbours around most of the decisions and practical work that this shared resource once required. Along with transformations in the town's economic base, new collective challenges require new expressions of community.

Notably, the state and market are each implicated in both the erosion and the rebuilding of community institutions responsible for managing and preserving Panajachel's common resources. The same formal structures of municipal government that in recent decades displaced community networks focused on the common waterworks are now, in new hands, actively supporting opportunities for strengthening networks of collective action at the neighbourhood scale and beyond. The expansion of tourism-related businesses, responsible for the transformation of Panajachel's traditional agricultural economy and associated changes in the town's social, religious, and ethnic composition, is eroding the traditional structures of Mayan civic-religious governance. At the same time, the welfare of the community broadly depends on preserving the basis of the tourism market. The business networks that have grown to protect this common interest include the largest hotel owners as well as street vendors and sand collectors with very limited resources.

On what does this reassertion of community depend? One often-repeated claim is that sustained patterns of collective action to maintain the commons tend to be associated with communities that are small, traditional, and homogeneous by virtue of ethnic, cultural, and religious identity. In our judgment, however, framing the issue in these terms misses the underlying basis for the creation and maintenance of the commons, which is not general identity but rather the strength of shared interests-things valued in common-even among people with diverse characteristics. Networks of community are based upon shared interests and the relationships that form around these. We do not mean to discount the importance of ethnic, religious, cultural, and economic diversification: in Panajachel, all of these make more challenging the task of identifying shared interests and sustaining relationships that span the various groups within the town. Homogeneity, however, cannot be considered a precondition for cooperative action to maintain the commons.

The key element that links the promising recent efforts by government, civil society, and business associations to address the problems of wastewater management in Panajachel is an explicit attempt to identify a common interest that bridges diverse social groups. Earlier attempts to address the problem were restricted to very limited social networks within the town. The few business leaders who recognised the problem in the early 1980s complained to local authorities but otherwise kept the conversation essentially among themselves. The mayor who commissioned the treatment plant was effective in engaging external finance and technical support but had neither the orientation nor the capacity to engage a public dialogue in the town itself. Likewise, the specialists contracted to design the plant were neither assigned nor prepared to consult broadly to understand the nature of the problem or explore alternative technological solutions with the local population. Even the government body specifically assigned to facilitate stakeholder collaboration to protect environment and livelihoods in the lake basin had been largely ineffective.

By contrast, the recent actions we have characterised as a reassertion of community have begun with conversations that engage a broad range of networks. These include networks defined by common economic interests, such as the larger tourism businesses that dominate the Grupo gestor, smaller business owners represented through the commerce association and the boat-owners association, as well as small merchants and wage workers. They include networks defined by place, representing the diversity of neighbourhoods within the town, which include poor and densely settled alleyways as well as groups of lakefront chalets with expansive gardens. They include networks, in particular the civic committees, formed previously in response to a shared interest in specific problems ranging from the need for a health centre to garbage collection to rebuilding the library. And they include networks of shared interest that extend well beyond the town itself, such as those focused on environment and sustainability at national and international scales, in which some of the larger NGOs, government agency staff, and some foreign and local business owners participate.

Underlying the apparently straightforward financial obstacle to operating and maintaining the Jucanya wastewater treatment plant is a challenge that runs much deeper, an initial failure to "build" the social and organisational foundations for project sustainability. This failure probably exacerbated conflicts in Panajachel after the plant's inauguration, but growing recognition of the critical importance of the problem to the town's social and economic welfare has subsequently opened avenues for public dialogue and community action. The theoretical implication of this analysis-the need to recognise the potential of state and market institutions as forces that may both erode and rebuild community-is matched by an equally important practical lesson. The planning, maintenance, and operation of public infrastructure projects need to be integrated into a process of community building by identifying shared interests and forming or strengthening the relationships needed for collective action.

Leaders of business, community organisations, and local government in Panajachel are facing this question: how can the social relationships and shared values to address wastewater pollution be cultivated once the physical infrastructure is already in place? Remarkably, even the engineer who designed the plant has concluded, reflecting on his experiences in Panajachel and with plants elsewhere in Latin America, that while selecting appropriate technology is important, social organisation must be considered the overriding determinant of a project's success. "The residents in the area of a water treatment plant have the political power to paralyse the project," he has written recently. "Rightly, this obliges the public sector (and private developers) to work better and to take account of citizens. There must be a process to educate and involve the community".

Even if judged by such criteria of consultation and participation, now broadly acknowledged at least in principle in the field of development planning and administration, the Jucanya wastewater treatment project falls far short. The project we have analysed lacked input from and consultation with community members even on narrow questions such as the proposed fee schedule. Indeed, even the contact made with local government was inadequate to plan appropriately for operation and maintenance of the facility, let alone to discuss the upstream questions related to choice of technology. There was also no systematic popular education or awareness campaign about the effects of wastewater pollution or the rationale for the project. In one respect, these facts are not surprising: even if development planners now commonly acknowledge the importance of awareness building, consultation, and community participation, the costs for these activities are rarely internalised into the budgets for infrastructure projects-and they ought to be.

Our argument, however, is more than a reaffirmation of the need for consultation or participation in the typical sense that these are employed as tools of development planning and project management. More fundamentally, what planners responsible for the wastewater treatment project lacked at the outset was an ethical commitment and practical capacities to participate in the making of community around shared interests defined and recognised by local actors.

For all the difficulties it entails in a context of rapid social and economic transformation, the longer-term prospects for addressing wastewater management in Panajachel depend on the ability of local residents to reassert community ownership and collective responsibility. The expressions of community emerging today-in neighbourhood committees, business associations, and municipal work projects-differ from prior patterns of collective work focused on the irrigation waterworks. But the challenges are in many ways parallel. The realignment of community institutions underway in Panajachel is far from settled. However, the moves toward reasserting community are significant, particularly in the midst of Guatemala's troubled postwar reconstruction, where the challenges of rebuilding social ties and faith in government remain huge.

Scholars and practitioners of development need to be able to critique-and hopefully avert-processes of community disembedding that threaten local livelihoods, and we also need to be able to identify and support processes of community building. The analytical dimension of this work requires an appreciation of the potential of both state and market forces as threats to as well as opportunities for a reassertion of community networks necessary to maintain the commons. The practical dimension of this work requires, in addition, a distinct ethical commitment that recognises the paramount role of community in social change, combined with capacities to act on that commitment.

Bacterial Pathogens Recovered from Vegetables Irrigated by Wastewater

An increase in consumption of fresh fruits and vegetables worldwide has been paralleled by an increase in the number of foodborne illnesses attributed to fresh products. Numerous reports have indicated that raw vegetables may harbour potential foodborne pathogens. In particular, tomatoes, cantaloupes, and sprouts have been linked to outbreaks of salmonellosis, and outbreaks of illnesses caused by Escherichia coli O157:H7 have been associated with melon, apple cider, lettuce, and radish sprouts. Moreover, coleslaw, cabbage, potatoes, radishes, bean sprouts, and cucumbers contaminated with Listeria monocytogenes have been linked to disease outbreaks, and salad vegetables also may be contaminated with Campylobacter.

Vegetables can become contaminated with pathogenic organisms during growth, harvest, postharvest handling, or distribution. Use of untreated wastewater in irrigation represents an important route for transmission of these pathogenic organisms. Raw vegetables are considered by some to represent an increased risk to public health when irrigation methods use untreated wastewater and no chemical treatments are employed to reduce the microbiological load on the raw product. In Morocco, vegetable products have been in great demand in recent years. Since the rate of precipitation has been very low during these last decades, wastewater is increasingly being used in agriculture. Little information is available on the number of human foodborne-illness outbreaks that have occurred from consumption of raw vegetables. The use of raw sewage to irrigate crops is an important mechanism that helps to propagate conditions conducive to cholera and typhoid fever. Increases in foodborne illnesses during the summer are not fully understood, although fresh produce likely plays a role since it is consumed in higher quantities during the summer. The study reported here investigated the occurrence of pathogenic bacteria in vegetables irrigated by untreated wastewater in Morocco. Irrigated vegetables do not undergo any control before being exposed in the markets, after which they may be eaten cooked or raw. The purpose of the study was 1) to determine the bacterial quality of vegetables irrigated with untreated wastewater, 2) to sensitize farmers to the dangers from use of untreated wastewater for irrigation, and 3) to elucidate the risk to Moroccan public health.

A total of 50 vegetable samples were procured for bacteriological examination. Vegetables of various types were obtained from several wastewater-irrigated agricultural regions in Morocco. Sampling was conducted from August 2002 to July 2004. The vegetable samples were collected in sterile polyethylene bags, and steps were taken to avoid contamination of the vegetables by soil or other contamination sources. Each sample was collected in triplicate to prevent sampling error. The vegetables were tomato, radish, cucumber, eggplant, potato, pepper, garden pea, gourd, zucchini, artichoke, broad bean, turnip, onion, French bean, and lettuce. All the samples were transported to the laboratory under low temperature (<7[degrees]C) and stored at 4[degrees]C until testing. They were analysed within 20 hours of sampling. Each sample was rinsed several times with sterile distilled water to eliminate the soil. Before analysis, 25 g of each sample was homogenised for two minutes with 225 mL of 0.1 percent sterile peptone water with a Model 400 Stomacher and serially diluted.

Bacteriological Analysis

Using the spread-plate technique and 100 [micro]L from the serial dilution, the authors prepared duplicate plates for the determination of aerobic plate counts (APC), Enterobactericiceae, fecal coliforms, total colilorms, Staphylococcus, and Streptococcus. Aerobic plate counts were made with plate count agar (Merck), and plates were incubated at 30[degrees]C for 48 hours. Then all colonies on plates were counted.

Enterococci counts were made with Slanetz and Bartley agar (Biokar). The plates were incubated at 37[degrees]C for 48 hours, and all typical colonies (pink or dark red with a narrow whitish border) were counted. For the coliform counts, violet red bile agar (from Merck) was used for direct plating, and plates were incubated at 37[degrees]C for 24 hours and 42[degrees]C for total coliforms and fecal coliforms, respectively.

Typical colonies were round, red to pink, 0.5 to 2 mm in diametre, and surrounded with a red-to-pink halo. Staphylococcus aureus counts were determined with Baird-Parker Agar (Difco) with egg yolk—tellurite emulsion, and plates were incubated at 37[degrees]C for 24 hours to 48 hours. Colonies selected from the agar surface were examined under microscope for Gram stain and were tested for catalase reaction and then for coagulase activity with plasma rabbit (Biokar). To isolate Salmonella spp., we pummeled a 25-g sample in a stomacher with 225 mL of buffered peptone water and pre-enriched the homogenate 37[degrees]C for 18 hours.

Figure: *For drinking water quality checks, microbiological analysis are done*

A 100-[micro]L sample was subcultured into 10 mL of Rappaport Vassiliadis Broth (Difco) and enriched at 41.5[degrees]C for 24 hours and 48 hours. One mL of the pre-enrichment broth was simultaneously inoculated into 10 mL of selenite cysteine broth and enriched at 37[degrees]C for 22 hours and 48 hours. Both enrichment broths were streaked onto xylose lysine deoxycholate agar (Merck) and Salmonella-Shigella agar, and incubated at 37[degrees]C for 22 hours. For selective plating, presumptive Salmonella colonies from selective plates were confirmed with the API 20E identification system (BioMerieux). The Enterobacteriaceae strain was isolated with Levine-EMB agar (Merck). The plates were incubated at 37[degrees]C for 18 hours, and colonies growing on the plates were examined under a microscope for Gram stains and tested for catalase and oxydase reactions. For identification of all strains, the API 20E identification system (BioMerieux) was used.

These analyses showed high aerobic-plate, total-coliform, fecal-coliform, and enterococci counts. Coagulase-positive Staphylococcus aureus was not detected in any samples. Citrobacter freundii and Enterobacter cloacae were recovered most frequently (from 28 percent of samples). Other Gramnegative bacteria that were frequently isolated were Escherichia coli (16 percent), Enterobacter sakazakii (12 percent), Klebsiella pneumoniae (17 percent), and Serratia liquejaciens (11 percent).

Foodborne diseases remain an important public health threat worldwide, and one of the most important food safety hazards is associated with raw vegetables. The large number of total microorganisms and fecal-contamination indicators (E. coli, coliform, and enterococci) detected in the vegetable samples we surveyed indicates a potential health hazard to consumers. Madden has discussed potential sources of microbial contamination of fresh fruit and vegetables during growth, harvest, distribution, and processing (1992). The bacteria that the authors found on samples belonged most frequently to the Citrobacter-Enterobacter-Serratia group of Enterobacteriaceae.

Although usually regarded as human pathogens, these members of Enterobacteriaceae family have also been recognised as inhabitants of soil and plants. Thus, vegetables may serve as a reservoir from which the bacteria named above can colonise and infect a susceptible host. In many countries, urban wastewater is used to irrigate agricultural land. This way of disposing of urban sewage water has several advantages. Wastewater contains a lot of nutrients, which increase crop yields without use of fertilizer. Furthermore, sewage water is an alternative water source in arid and semi-arid areas where water is scarce. Some disadvantages are that wastewater can contain heavy metals, organic compounds, and a wide spectrum of enteric pathogens that may have a negative impact on the environment and human health.

The study reported here demonstrates that a potential for disease transmission exists when wastewater is used for irrigation. Pathogens that have been transported by wastewater can survive in soil or on crops. The actual risk of disease transmission, however, is related to whether this survival time is long enough to allow transmission to a susceptible host. The crop and the field are the link between the pathogen in the wastewater and the potential for infection. The factors controlling transmission of disease are agronomic; examples of such factors are the crop grown, the irrigation method used to apply wastewater, and cultural and harvesting practices.

Consumption of salad irrigated by wastewater has been found to be responsible for shigellosis in England. Numerous opportunities exist for attachment and penetration of pathogenic bacteria into lettuce in the field, as well as during harvesting, processing, and marketing, especially when a contaminated product is exposed to water or is damaged. Guo and co-authors have demonstrated that soil and water are potential reservoirs from which Salmonella can contaminate tomatoes. The pathogen can survive in most soils in high numbers for at least 45 days and can infiltrate the tissues of tomatoes during contact with inoculated soil. This mechanism may explain the discovery of Salmonella arizonae on tomatoes in the investigation reported here. The hydroponics system used in another study provided a controlled environment in which to study the possible association of Salmonellae with aerial tissues, with minimal concerns about environmental contamination or temperature fluctuation.

That study provided evidence that Salmonellae can be transported from an inoculated nutrient solution to the hypocotyls, cotyledons, stems, and leaves of young tomato plants. In addition, the work of Guo and co-authors (2001) revealed the ability of Salmonella to survive on or in tomato fruits throughout the course of plant growth, flowering, fruit development, and fruit maturation (2001). Another study showed that Escherichia coli O157:H7 may be present not only on outer surfaces, but also in the inner tissues and stomata of cotyledons of radish sprouts grown from seeds experimentally contaminated with the bacterium (Itoh et al., 1998). Indeed, this mechanism can be inferred from the fact that enteric pathogens were isolated from inside vegetables during our investigation.

Numerous foodborne diseases caused by Salmonella enterica and Escherichia coli O157:H7 have been associated with contaminated alfalfa, clover, and bean sprouts. Constant moisture, nutrients released by the sprouting seeds, and warm temperatures are conducive to the growth of human bacterial pathogens such as S. enterica and E. coli O157:H7. The fact that some bacteria and not others are present can be explained by a difference in their capacity to attach to plants. For example, Salmonella enterica attaches as well as plant-associated bacteria and significantly better than E. coli to alfalfa sprouts.

The authors' study demonstrated that the vegetables analysed may be an important source of foodborne bacterial-illness outbreaks in humans, especially when the vegetables are consumed raw. In addition, vegetables of these types can also be a vehicle for Giardia cysts and Ascaris eggs. To protect public health, the use of raw sewage in the irrigation of the vegetable culture must be prohibited. So it is necessary to consider the generalised program of wastewater treatment in Morocco. Moreover, the requirements for treated wastewater must respect the sanitary standards for agricultural reuse. There is an urgent need for development and validation of standard methods of eliminating the pathogenic microorganisms from raw vegetables.

5

Waste Discharges and their Effects on Receiving Waters

A complete description of the physical, chemical, and biological aspects of water quality, as a function of natural and man-made factors, and the technological measures available for changing water quality can easily fill several extensive volumes. The aim is to provide a background for consideration of the economic and institutional aspects of water quality management. Because each type of receiving water has characteristics that, to some degree, result in differential impacts of the same waste discharge on water quality, no attempt at complete coverage is made. The relation of these various methods of region-wide water quality and water resources management will also be indicated.

Major Types of Wastes and their Sources

The many substances having an impact on water quality that enter surface watercourses and groundwater aquifers as a result of man's domestic, industrial, agricultural, and recreational activities can be grouped in various ways. One common classification differentiates between physical substances (e.g., radioactive, inorganic, organic) and biological substances (e.g., bacteria and viruses). In addition, the quality of a waste discharge can be measured in various ways. Some of the common measures are biochemical oxygen demand (BOD), chemical oxygen demand (COD), alkalinity, acidity, colour, hardness, taste, odor, turbidity, and electrical conductance. For the purposes of this book, however, the most useful classification of wastes distinguishes between nondegradable and degradable substances, based on their behaviour in receiving waters. Nondegradable wastes are usually diluted and may be changed in form, but they are not appreciably reduced in weight in the receiving water. Degradable wastes are reduced in weight by the biological, physical, and chemical processes which occur in natural waters.

Nondegradable Wastes

The sources of nondegradable substances mainly inorganic chemicals, such as chlorides, synthetic organic chemicals, and inorganic suspended solids are many and widespread. Industrial waste discharges frequently contain inorganic or metallic salts, synthetic organic chemicals, and other materials which may be toxic or corrosive, may impart colour or taste to the water, and or may produce odors. Domestic water use results in a small increase in the content of chlorides and other dissolved salts. The return flow from irrigation is generally significantly higher in dissolved salts than the applied irrigation water. The discharge of copper, zinc, uranium, and other compounds in drainage from mine tailings can have very adverse impacts on water quality, and acid mine drainage is an acute problem in some of the major coal-producing areas. In the heavily industrialised sections of the country and in the major irrigated areas of the West, industrial operations and irrigated agriculture may be major sources of chlorides and other dissolved salts.

Natural geologic formations, such as salt deposits, may result in high concentrations of chlorides and or dissolved solids in watercourses. Another instance of deterioration of water quality by a natural source is from seawater intrusion in groundwater aquifers near coastal areas (following a change in the hydraulic gradient because of pumping from the aquifers). Similarly, the quality of water in inland groundwater aquifers can be adversely affected by excessive pumping where there are hydraulic connections with brackish or saline aquifers.

Suspended materials, which can be considered nondegradable in the sense that the), are essentially not changed in weight in receiving waters, consist of sediment from natural and accelerated erosion from land surfaces and stream channels and colloidal material from clay deposits and in domestic and industrial waste discharges. Suspended sediment usually settles out, but colloidal matter does not, except under unusual circumstances. Both suspended sediment and colloidal matter cause turbidity in surface waters. This not only makes the water less attractive, but can damage fish life. Turbidity may also inhibit the growth of algae and other aquatic plants, which may or my not be undesirable.

Agricultural activities generate both nondegradable and degradable wastes. Storm runoff from agricultural land carries silt, clay, fertilizers, and weedicides, as well as organic material and bacteria, into watercourses. In contrast to the sources of wastes described above, which result in waste discharges concentrated at one or a few points, agricultural activities generally result in dispersed waste loads, except in the case of feed lot operations. In predominantly agricultural regions, for example the Missouri Basin, these dispersed waste loads may be the major sources of nitrogen, phosphate, and pesticides, as well as organic wastes.

Actually the classification into degradable and nondegradable waste loads is a simplification. Some types of substances, such as radioactive materials and some of the organic chemicals, are degradable to some degree. Viruses also appear to be in the in-between category. Some radioactive wastes decay rapidly; others extremely slowly. Some of the synthetic organic compounds, such as alkyl benzene sulfonate (ABS) detergents, are not strictly nondegradable, but they resist attack by stream biota. On the other hand, the new detergents based on linear alkylate sulfonate (LAS) are degradable, both in streams and other watercourses and in waste treatment plants.

One other point with respect to nondegradable wastes merits mention, namely, the problem of identifying the substances in the receiving water. With respect to many of the complex industrial, agricultural, and pharmaceutical chemical compounds, not only have techniques not yet been developed to identify them, but the concentrations involved are often on the order of a few parts per billion, rather than parts per million as with chlorides or suspended sediment. Further, the extent to which there are synergistic or antagonistic reactions is not fully understood as yet. Viruses also present identification problems because they can exist in a dormant condition outside living organisms and be reactivated under proper environmental conditions.

Degradable Wastes

Degradable wastes— the substances that are reduced in quantity by the biological, chemical, and physical phenomena characteristic of natural waters include organic wastes of various kinds from domestic and industrial operations, bacteria, and thermal discharges. The most widespread source of degradable organic materials is domestic sewage. In terms of total quantity, industrial discharges represent the largest source. The highly unstable, putrescible, organic waste can be converted to stable inorganic materials (bicarbonates, nitrates, sulphates, and phosphates) by the bacteria and other organisms found in natural water bodies.

If the water is not too heavily loaded, the process, commonly but somewhat deceptively known as self-purification, will proceed aerobically by the action of bacteria utilising free oxygen and will not produce offensive odors. If the receiving waters are loaded excessively, the process of degradation will proceed anaerobically by the action of bacteria not utilising free oxygen, and hydrogen sulfide and other gases will be produced. Both of these processes are utilised in waste treatment plants, which, in essence, systematize, control, and accelerate the processes that take place in natural waters.

Figure: *To handle degradable and non degradable wastes and Sanitation.*

Bacteria are omnipresent in natural waters. Except for a few types, i.e., the pathogens, the vast majority are beneficial. The primary waste discharges that are sources of bacteria are from activities of warmblooded animals. The coliform count is the most common index of such waste discharges. Total bacteria count is also used as an indicator of water quality in relation to such discharges. Both coliform count and total bacteria count serve as crude proxies for the actual substances of concern, namely, the bacteria that cause infectious diseases primarily typhoid, dysentery, and cholera. Bacteria may be considered degradable since they tend to die off rather quickly after leaving the body.

Thermal loads are classed as degradable because the heat is dissipated in receiving waters, primarily by evapouration in surface water, and by some combination of evapouration and conduction in groundwater. The primary sources of thermal loads are the generation of electrical energy and the cooling operations in the petroleum refining, pulp and paper, iron and steel, and chemical manufacturing industries. Thermal discharges are particularly troublesome when industrial operations are concentrated along a stream or an estuary.

Effects of Waste Discharges on Receiving Waters

Nondegradable Wastes

Given the weight of a waste in a discharge and the volume of the discharge, the concentration of a nondegradable waste over time in a stream or lake is relatively easy to predict because dilution is the dominant process when the substance is in solution. Prediction is more difficult for estuaries because of the complex hydraulic characteristics resulting from tidal action and topography. Inorganic chemicals may precipitate out in lakes, reservoirs, and, under certain conditions, in flowing streams.

Suspended sediment will settle out to some degree in receiving waters, particularly in sections where there is a major decrease in stream velocity, although the spatial pattern of deposition cannot always be accurately predicted. For a nondegradable waste discharged to a groundwater aquifer, predicting the concentration over time at various distances from the original discharge point is complicated by variations in transmissibility, both vertically and horizontally, and by the physical processes (such as adsorption) and chemical reactions which are possible. In all of these situations the weight of the discharged substance remains essentially unchanged, although the form may be modified in the receiving waters.

It should be emphasized that, while the prediction of concentration of a nondegradable waste in a specified water body is relatively easy for a known weight of waste and given volume of receiving water, prediction of the actual time pattern of concentration resulting from a waste discharge is far less simple. This is true for two reasons. First, with respect to the receiving water, there is uncertainty about the quantity of water available for dilution, i.e., hydrologic uncertainty. Hydrologic events are stochastic variables, hence the quantity of water available is known only in terms of a probability distribution. Second, where the source of the nondegradable substance is an industrial operation, there often is significant variation from day to day (and even within the day) in the quantity of waste discharged. Despite these difficulties, procedures have been developed which enable reasonably accurate prediction of time patterns of concentrations resulting from defined time patterns of discharge of nondegradable wastes.

Degradable Wastes

Predicting the time pattern and spatial pattern of concentration of wastes presents considerably more difficult technical problems when the wastes are degradable than when they are nondegradable. This is particularly true for organic wastes, but it holds to a lesser degree for thermal discharges, for certain types of radioactive wastes, and for those synthetic organic pesticides which decompose to some degree. With respect to organic and thermal waste discharges, the primary concern is not the fate of the wastes *per se*, but the effect of their degradation on water quality, in particular the dissolved oxygen concentration in the receiving waters. Because of the importance of organic waste discharges in relation to total waste discharges and because of their widespread occurrence, we turn to a relatively detailed discussion of such wastes. An understanding of the phenomena involved is essential for the discussion of the economic, technological, and institutional aspects of water quality management in the remainder of the book. A measure of organic waste load is biochemical oxygen demand (BOD), which indicates the amount of oxygen drawn upon in the process of decomposition of the waste. The amount of oxygen demanded and the rate at which it is drawn upon are functions of the type and quantity of the waste and of other factors, among which the most important are the chemical characteristics and the temperature of the receiving water. Toxic substances, for example, may appreciably reduce the rate of decomposition by inhibiting bacterial action, and in extreme cases of toxic concentration a body of water may become bacteriologically "dead." At higher temperatures, when the oxygen saturation level of water is relatively low, bacterial action is accelerated, wastes are degraded more rapidly, and dissolved oxygen in the water is drawn upon more rapidly.

The imbalance between available oxygen and oxygen demand may proceed to the point where septic (anaerobic) conditions result. Such conditions are most likely to occur in summer and early fall when streamflows tend to be low, and temperatures high.

The rate at which BOD is exerted combined with the rate at which oxygen is restored determines the level of dissolved oxygen (D.O.). In flowing water the combined effect of an organic waste discharged at a specific location and reaeration in the stream results first in a decrease and then in an increase in D.O. as the waste is carried or moved downstream. This phenomenon is illustrated by a characteristic curve known as the "oxygen sag." The variables affecting the shape of the curve can be described by standard equations, the development.

Other things being equal, factors that reduce the rate of BOD lengthen and flatten the oxygen sag, while those that accelerate BOD have the reverse effect. The shape of the oxygen sag is also affected by the rate of reaeration, which depends largely on channel characteristics as they affect turbulence and the area of the air-water interface, the net photosynthetic oxygen production, and the velocity of streamflow.

The role of photosynthesis, as a producer of oxygen, in the dissolved oxygen balance of receiving waters is a complex one. Photosynthesis depends on light and nutrients available, stream velocity, amounts and types of waste loads, water depth, and water temperature. Most of the photosynthesis in water bodies is performed by aquatic green plants called algae, which range in size from tiny, single-cell plants to various types up to several feet in length.

The contribution of photosynthesis to the oxygen content is complicated by the problem of respiration. An important factor in the assessment of this contribution is the P/R ratio, the ratio between oxygen production and community respiration, where community includes aquatic animals, plants, and bacteria. Directly downstream from the discharge of an organic load the P/R ratio is usually very small, i.e., on the order of 0.01. The peak production of oxygen occurs in the early portion of the "zone of recovery", where the P/R ratio may be on the order of 3.0 or more. Downstream a significant distance from the waste discharge, as the dissolved oxygen content of the stream approaches the level prior to the waste discharge, the P/R ratio tends to approach unity. The wide variation in the P/R ratio which has been found in various streams complicates the analysis of the impact of organic waste loads on dissolved oxygen.

A further complicating factor with respect to organic waste loads and photosynthesis involves the diurnal variation in dissolved oxygen. The diurnal variation may be of particular importance for certain types of water uses, such as fisheries. Thus the analysis of water quality based solely on longer time periods, i.e., weeks or months, may in certain instances be inadequate as a basis for optimal water quality management.

The low point on the oxygen sag has been a governing factor in planning or determining the size or capacity and the design of waste treatment plants. Of course, "point" in this context actually means a particular reach or an area of a given stream, estuary, or lake. Sanitary engineering practice has typically been to calculate the oxygen sag at a given "low" level of streamflow and to determine the capacity and design of a waste treatment plant which would result in a specified level of dissolved oxygen in the critical reach of the oxygen sag. The stochastic nature of streamflow and its effect on the variation in assimilative capacity of the stream, and hence on the possibilities for more flexible treatment facilities, have generally been ignored in this planning process, at least until recently.

The equations detailed were first devised to describe the behaviour of D.O. in a single reach of stream subject to a single waste discharge. They can, however, be applied to multiple reaches, since the deficit at the downstream end of one reach is also the initial deficit (D_a) at the upstream end of the next reach. Such straightforward use of the equations is possible only under highly simplified conditions, including a constant oxygen saturation along the length of the stream. The basic equations describing D.O. behaviour have been elaborated in a number of ways to handle more complex situations including both multiple reaches and multiple points of waste discharge.

Consideration of waste disposal and water quality management on a regional basis requires methods for tracing the effect of waste discharges on long reaches of streams, in lakes, and throughout estuaries and groundwater basins. Where wastes are discharged to flowing streams from closely spaced outfalls, the level of D.O. at any given point is a function of the rate of reaeration and the exertion of BOD from the various sources, and of course the factors that affect both. (The same is true for lakes and estuaries.) At a given downstream point the BOD from the farthest upstream sources may be almost completely exhausted, while that from less distant sources will still be exerted at a high rate. A dissolved oxygen profile for a long stretch of such a stream will show deep valleys below large waste discharges or closely spaced small discharges and peaks at or near the D.O. saturation level at other locations.

Among the complexities that beset the calculation of oxygen levels, especially for long reaches of a stream, is the fact that BOD proceeds in two stages. When an organic waste is discharged into a stream where dissolved oxygen is near saturation level, there is an immediate and major draft upon the D.O. content of the stream, as the organic wastes are degraded by bacterial action. Thereafter the D.O. level tends to recover. Farther downstream, corresponding roughly to a travel time of five to seven days, a "second stage" BOD occurs, as the nitrogen embodied in the organic wastes is converted to nitrite and then to nitrate by aerobic nitrifying bacteria. The second stage is more diffuse and does not by itself reduce the D.O. to as low a level as the first stage does.

Tracing the fate of an organic waste load in an estuary is even more difficult because the estuarine environment is complicated by tidal action, the large expanse of air-water interface, more complex hydraulic characteristics, and generally complicated patterns of inflows. In addition, photosynthesis may play an even larger role in the oxygen balance of an estuary than of a flowing stream.

Despite these complexities, significant advances have been made in developing models which interrelate the many variables involved, such as streamflow and temperature with less or more consideration of the stochastic nature of both and multiple waste loads, and enable prediction of regional profiles of D.O., temperature, and other water quality characteristics. These analytical models, virtually all of which depend on the use of computers, have provided the bases for economic optimisation studies of water quality management systems.

To cite one example, Thomann developed a mathematical representation of the Delaware estuary water quality system, composed of two subsystems: (1) the biochemical oxygen demand subsystem; and (2) the dissolved oxygen subsystem. For each of the sections into which the estuary was divided, mass balance equations were written for both subsystems, assuming steady-state systems, i.e., invariant with time. These consisted of two linear differential equations based on the hydrological, physical, and biochemical characteristics in each section. The result was two sets of thirty simultaneous equations.

By matrix manipulation a set of transfer functions was obtained from the coefficients of these equations. This set of transfer relationships yielded the transformation of a waste load input in any specific section to the dissolved oxygen level in any other section. Since all of the equations in the system are linear, the solutions could be linearly superimposed. Thus, the total effect at any section was found by adding the effects in that section caused by inputs at any number of sections in the estuary.

Then the assumptions of a stationary system were removed, i.e., the parameters in the equations were permitted to vary with time, and a digital computer was used to solve the equations. Comparisons between the model results and actual data provided verification of the model and simultaneously the more accurate specification of the variables in the equations. The verified model was the basis for analyses of various combinations of inflow to the estuary, waste discharges, and additions of supplemental oxygen, and variations in reaeration and degradation rates.

The basic phenomena involved in the dissolved oxygen-organic waste load system of the Delaware estuary were reaeration, degradation, and tidal diffusion. By modifying the model, e.g., eliminating the degradation process, the behaviour of nondegradable wastes, such as chlorides, could be investigated.

The degradation of organic wastes discharged to groundwater occurs primarily by anaerobic decomposition. Detailed studies of the dispersion and degradation of organic wastes discharged to porous media have been made. Lakes are another special case. Being more stationary than streams or estuaries, they are naturally more susceptible to eutrophication (nutrient buildup).

This process is speeded up by the discharge of organic wastes, even residual wastes after significant levels of treatment. Nutrient enrichment tends to be accompanied by oxygen deficiencies and can have a serious effect on water quality, as evidenced by conditions in Lake Tahoe (California-Nevada), and Lake Erie. Eutrophication, the complex circulation patterns of large lakes, stratification and temperature inversions, all combine to make the analysis of the effect of organic waste discharges on water quality in lakes very difficult.

The attention presently being devoted to studies of lacustrine environments suggests that progress should be forthcoming in developing methods for simulating the behaviour of the many interrelated variables in such environments.

The residual products of organic waste degradation in water bodies are plant nutrients, i.e., nitrogen, phosphorus, carbon. Under propitious conditions of water temperature and water clarity, these plant nutrients contribute to algae growth, which in turn affects the oxygen balance in the receiving water. If algae are present in large quantities, they may affect the appearance, taste, and odor of the water.

As can be deduced from the previous discussion, degradable wastes in the form of thermal discharges are of importance primarily because of their effect on water temperature, which in turn affects both the rate of reaeration and the rate of decomposition of organic wastes, and hence affects D.O. As in the case of organic waste discharges, methods have been developed to predict the dissipation of thermal discharges in water bodies.

While simple classification systems for types of wastes and types of receiving waters comprise a convenient point of departure for discussing the technical background for water quality management, they obscure, to some extent, the highly complex phenomena and interrelationships which are involved. A vast number of substances enter watercourses. Some result from man's activities and others are of natural origin.

Their effects are interrelated in many complicated ways. For example, water clarity is an important precondition for the production of algae. Thus, streams with high turbidity are not likely to have the taste and odor problems that often accompany prolific algae growth. Consequently, when measures are taken to reduce turbidity, thereby increasing the clarity of water, algae production may increase with resulting taste and odor problems that previously did not exist.

Lakes, reservoirs, and estuaries, like flowing streams, are subject to the deposition of sludge banks, which if suddenly dispersed produce a "shock load" of oxygen demand. But these bodies of water also have special problems of their own. Stratification, the formation in essentially stationary bodies of water of thermal layers which prevent deep circulation, complicates the analysis of water quality in lakes and reservoirs.

Plant nutrients also take on special significance because they tend to accumulate over time in relatively stationary bodies of water and contribute to the depletion of oxygen in the lower levels. In a tidal estuary, wastes may be moved back and forth within the estuary for long periods of time rather than being dispersed into the ocean. Thus, residual wastes may build up with resulting water quality deterioration.

Much has been done to develop generalisations about the effect of waste discharges on the quality of various types of receiving water. However, the accuracy and precision are less and the uncertainty is greater in relation to predicted behaviour of these phenomena than in many fields of scientific and engineering analysis.

Therefore, it is important that any estimates of behaviour based upon simplified models be checked with empirical data. Nevertheless, much progress has been made in developing mathematical models of waste transport, dilution, and degradation processes in receiving waters.

Several models are now operational and capable of predicting, with a reasonable degree of accuracy, the sequence of results over time and space of the discharge of wastes.

The Assimilative Capacity of Receiving Waters

The quality of water bodies can be improved by reducing the waste loads discharged to them, by increasing or making more effective use of their assimilative capacity, and by combinations of the two. If the focus is on water quality management in streams, and if there is no problem with floating materials, reducing waste loads and changing assimilative capacity are technical substitutes for one another. Assimilative capacity can be increased by altering the time patterns of streamflows, by altering the time pattern and/ or locations of waste discharges, by artificial reaeration, or by some combination of these measures. If the assimilative capacity is known, intake water treatment can be a partial substitute for waste load reduction. Indeed, if water were not useful, offensive, or altered in quality while *in situ*, intake water treatment would be technically interchangeable with waste load reduction. Flowing streams, of course, are much more amenable to procedures for changing assimilative capacity than are groundwater aquifers or even reservoirs and lakes.

Improving the water quality of receiving waters by reducing waste loads can be accomplished in two broad ways: first, by reducing the generation of wastes; and second, by modifying the residual wastes. Within these two broad categories a variety of techniques are available. The dividing lines between methods that reduce waste generation and those that affect the residual wastes are not always as clearcut as implied in the table. Materials recovery, for example, is in a sense a change in production process because it involves a change in the mix of raw product inputs; process change may be necessary before in-plant water recirculation is possible; and product output changes may require some changes in production processes. Nevertheless, the distinctions are worth preserving as a basis for discussion. For example, following generation, waste materials may be recovered, used for the production of byproducts, reduced by waste treatment, temporarily stored prior to discharge, conveyed to another water user, or discharged in various possible ways into one or more receiving water bodies.

Thus, "municipality" could be substituted for "production process". This would require only slight modification of the diagram, i.e., elimination of the line to "materials recovery, etc." and of the line indicating direct recirculation. Industrial wastes play such a large role in water quality management that an understanding of the factors that influence the generation of these wastes is extremely important. This is not to say that the magnitude of wastes generated by domestic and commercial users is insignificant. On the contrary, with expanding population, particularly in urban areas, the absolute quantity of domestic wastes generated is bound to increase. And with the widespread adoption of garbage and paper grinders, there may also be an increase in the generation of wastes per capita within the home.

***Table:** Methods for Improving the Quality of Receiving Waters*

Methods for Reducing Waste Discharges

Methods for reducing wastes generation

1. Change in type of raw material inputs
2. Change in production process
3. Change in product outputs
4. In-plant recirculation of water.

Methods for reducing wastes after generation

1. Materials recovery
2. By-product production
3. Waste treatment
4. Effluent reuse (including groundwaterrecharge, wastewater reclamation or renovation).

Methods for Increasing or Making Better use of Assimilative Capacity

1. Addition of dilution water
2. Multiple outlets from reservoirs
3. Reservoir mixing
4. Reaeration of streams
5. Saltwater barriers
6. Effluent redistribution (including regulated discharge) .

But industrial waste discharges currently account for the greater share of the residual materials discharged to receiving waters in the United States, and their control is a complex, multi-faceted process. In contrast, treatment is the only general method for reducing waste discharges from households, barring significant changes in consumption patterns. Accordingly, the primary focus is on industrial wastes.

Improving Water Quality By Reducing Waste Discharges

Factors Reducing Wastes Generation

The generation of industrial wastes per unit of raw product processed or per unit of final product is a function primarily of the type of raw materials used in the production process, the technology of the production process, the product mix, and sometimes the extent of in-plant water recirculation. Thus there is a wide range in the wastes generated per unit by different industries, and also by different plants within a single industry. To illustrate, a German study reported ratios of high to low waste discharge (in pounds of BOD per unit of product or raw product processed) within several industries as follows: about 25 for paper mills, 2.5 in malt factories, 6 in starch factories, 50 in pork slaughterhouses, about 10 in tanneries, and about 20 in textile factories. Similar variations occur in U.S. industries.

Shifting to a different grade or type of raw product is one way of reducing waste loads. The sulfur content of crude oil has an important bearing on a petroleum refinery's waste. In pulp and paper, the type of wood used is significant. In fruit and vegetable canning, the size, shape, per cent of solids, and resistance to damage and bruising in harvesting and processing the raw product affect the waste load per ton of raw product processed.

Changes in production processes, however, have the major impact on wastes generation, although it should be emphasized that technological changes have not been, and are not usually, instituted because of water quality problems. In fact, most changes in production technology have been stimulated by factors unrelated to water problems, and have been developed without explicit consideration for their effects on water quality. On the other hand, where various stimuli to management, such as sewer charges, have been used, they have often resulted in process modifications. As more systematic means are developed to bring to bear on industry the broader social costs associated with the discharge of waste materials into the environment, the generation and control of wastes will receive more prominent consideration in process design, and perhaps even in product development. Studies of several industries make clear that even comparatively small design changes can have a substantial effect on the generation of wastes.

Table: *Production of Wood Pulp in the United States, by Process and Year (per cent of total production)*

Process	*1940*	*1945*	*1950*	*1955*	*1960*	*1965**
Mechanical	18.2	18.0	14.9	13.2	13.0	11.8
Unbleached sulfite	11.1	8.0	5.0	3.1	2.1	1.5
Bleached sulfite	18.0	15.2	14.2	13.9	12.6	11.4
Unbleached sulfate	35.3	35.6	38.4	37.0	34.3	35.4
Bleached sulfate	6.5	8.4	12.1	17.5	23.3	26.3
Semi-chemical			4.6	6.8	7.9	8.7
Soda	5.9	4.2	3.5	2.1	1.7	0.7
All other	4.9	10.6	7.2	6.4	5.2	4.4
Total	99.9	100.0	99.9	100.0	100.1	100.2

Source: American Paper and Pulp Association, *Statistics of Paper-1964*, p. 10, and American Paper Institute, *Statistics of Paper, 1966 Supplement, 1966*, p. 2.

* = Preliminary.

Includes special alpha and dissolving grades.

Includes semi-bleached sulfate.

Reported in "all other" for 1940 and 1945.

Total production increased from 8.96 million short tons in 1940 to 33.3 million short tons in 1965.

In the pulp and paper industry the proportions of the total output produced by different processes have shifted significantly over the past two decades. The largest and most significant shift has been from the sulfite process to the sulfate process, which, by making the recovery of chemicals for internal reuse economic, can reduce the waste load in terms of pounds of BOD per ton of product to about 5 or 10 per cent of the previous level. The development of save-alls to recover waste fibers which formerly were simply discharged into water courses was stimulated by the increase in the value of fiber. Their use has resulted in a significant reduction in terms of pounds of both BOD and suspended solids per ton. Both black liquor recovery systems and save-alls were stimulated originally by the economic incentive to save materials useful in the production process, rather than by waste problems.

Other possibilities for process changes which will reduce wastes generated appear to be on the horizon. One of these is a "dry" papermaking process that will use considerably less water as a defibrating and distributing agent in the paper-making process. A particularly difficult waste problem in the steel industry for many years has been disposal of the "waste pickle liquor" generated when steel is pickled with sulfuric acid. Attempts to develop an economically feasible sulfuric acid recovery process have met with little success under U.S. conditions. Another solution is to shift to hydrochloric acid (HCI) in the pickling process itself. A recent report indicates that the acid losses in the effluent are reduced by this process change from about 13 pounds of sulfuric acid per ton of steel pickled to about 0.2 pound of hydrochloric acid per ton of steel. Hydrochloric acid offers other advantages as well a high-quality product (in terms of brightness of surface) and faster production. In the beet sugar industry the waste load generated in pounds of BOD per ton of beets processed has been reduced greatly in the last two decades by comparatively simple and economical changes in processes.

The substitution of drying of beet pulp for storage of wet beet pulp in silos and the use of Steffens waste for the production of by-products, or the complete elimination of Steffens processing can reduce BOD generation by about 60 per cent. Converting to pulp drying has another advantage if there is a market nearby, because dry pulp brings a greater return than wet pulp. The other process change, i.e., a shift from cell-type to continuous diffusers, is integrally related to recirculation of screen and press water. This change reduces the BOD generated by about 10 per cent.

Specific cases, as well as general industry examples, can be cited. For example, one chemical plant over a fifteen-year period doubled the dollar value of production and at the same time reduced the BOD and phenolic waste loads per day by over 90 per cent. The waste reductions were accomplished primarily by process modification, in many cases by simple changes such as increasing the temperature at which a reaction was taking place and changing a polymerization catalyst.

Table: *Wastes Generated in Beet Sugar Plants, by Process1 (pounds of BOD per ton of beets processed)*

Operation	*Potential BOD generation*	*After process changes*	*After recirculation associated with process changes*
Flume water	4.5	4.5	4.5
Screen water	2.52	2.53	0
Press water	2.6	2.63	0
Silo drainage	12.3	04	0
Lime cake slurry	6.5	6.5	6.5
Condenser water	0.7	0.7	0.7
Steffens waste	10.4	05	0
Total	39.5	16.8	11.7

Source: G. 0. G. Löf and A. V. Kneese, *The Economics of Water Utilisation in the Beet Sugar Industry* (Resources for the Future, Inc., 1968).

Based on typical 2,700 tons per plant day.

Average of BOD waste generation in batch diffusers and continuous diffusers.

Assuming conversion to continuous diffusers but without recirculation of screen and press water.

Assuming pulp dryer installation.

Assuming discontinuance of Steffens processing or the complete use of Steffens waste in by-product manufacture.

In the production of potato flakes, the processing of raw potatoes by steam peeling results in a BOD waste with a population equivalent of about 200 per ton in contrast to about 420 per ton using caustic peeling. Where cooling is a major component of an industrial process, changes in cooling methods are changes in production processes. The adoption of air cooling, as in some chemical plants and petroleum refineries, will reduce the thermal waste load generated in the production process.

It is interesting to note that air cooling may be the rational economic choice even where water is relatively inexpensive in terms of both supply and waste disposal problems. The cost associated with circulating a large quantity of air must be compared with the costs of high head pumps, pumping energy, and pipelines required to circulate water, plus the costs associated with water intake and wastewater disposal. There is obviously no waste to be disposed of from air cooling units. In addition to the lower investment costs with air cooling, there are operating savings from less equipment maintenance and improved continuity of operation.

In many cases it is difficult to separate changes in production processes from changes in products. Change in consumer demand over time is one stimulus for development of new production processes. Several examples illustrate the nature of product changes, with their implications for changes in wastes generation. In petroleum refining the product mix has shifted significantly over time toward the production of more gasoline per barrel of crude throughput. Continued changes in product mix are in prospect. For example, by 1985 the demand for jet-fuel will require about 16 per cent of all crude runs in the United States, in contrast to only about 3 per cent in 1965. This will require additional use of hydro cracking with consequent changes in waste generation per barrel.

In pulp and paper manufacture, there has been a steady shift toward a wider variety of products, such as colored paper and coated products of various types. Many of the coatings and sizings which have been developed to provide particular characteristics in the final products make reuse of the waste fiber more difficult or impossible, with a consequent increase in the waste load generated per ton of product. There also has been an increase in the proportion of pulp output which is bleached, from about 25 per cent in 1940 to almost 40 per cent in 1965. Bleaching increases the waste load generated in pounds of BOD per ton by about 25 per cent.

An example of a significant change in waste load through product change is the shift from hard detergents to soft detergents which are readily degraded by the biota in natural watercourses and in waste treatment plants. The change took place in Germany in October, 1964, and in the United States in the summer of 1965. Even so, soft detergents are still "under fire" because of their phosphate content, which contributes to algae growth in streams and lakes, and research is now oriented toward finding a substitute for phosphate.

Process and product changes have obviously had different effects on wastes generation. However, the dominant tendency has been toward a reduction in wastes generated per unit of raw or finished product. Two points are important:

(1) most of the changes in production processes and product mixes have occurred independently of problems of water quality management (and of air and solid wastes management);

(2) there are usually several ways at least of producing a given product mix, so that there are possibilities for modifications of production processes to reduce the wastes generated per unit.

One other type of in-plant change that can reduce waste generation is in-plant recirculation of water. As the extent of water recirculation in a plant increases, a decrease may occur in the waste load generated per unit, e.g., pounds of BOD and/or suspended solids per ton. This has occurred in the beet sugar industry and in the canning of some fruits. In the latter, where flume water is recirculated, less sugar is dissolved from the fruit in processing, thereby decreasing the BOD load per ton.

The separation of the effect of water recirculation *per se* on waste generation from the effect of changes in production processes and from materials recovery is not clearcut. In-plant recirculation of water may require modifications of production processes. Conversely, some production processes enable recirculation of water, as noted in the previous description of process changes in the manufacture of beet sugar. It is clear, however, that water recirculation in some cases has significantly reduced waste generation per unit.

Methods for Reducing Wastes after Generation

Materials recovery and by-product production are two methods for reducing wastes after generation. They form a logical pair, because the types of processes are often very similar. We differentiate between them on the basis of the final destination of their outputs. In materials recovery the output is reused within the same production unit as an input to the production process. By-product production yields consumption goods or intermediate goods used in other production processes.

Examples of materials recovery are to be found in almost all, if not all, of the heavy water-using or water-polluting industries. Only a few will be mentioned here for illustration. The recovery of chemicals in black liquor recovery systems in the production of sulfate pulp and the utilisation of save-alls in paper production have been mentioned previously. In the manufacture of synthetic phenol by the sulfonation process, liquid wastes have been essentially eliminated by process engineering, and the value of recovered materials is reported to exceed recovery costs.

In steel production, sedimentation to eliminate mill scale from waste discharges has resulted in the recovery of iron that can be used in the production process. Plating compounds, a potentially troublesome waste load, can be recovered from rinse water by evapouration. This procedure concentrates the rinse water sufficiently for reuse in the plating bath, which not only conserves valuable plating compounds but also significantly reduces waste and thereby the outlay for neutralizing chemicals. One plant anticipates an annual net saving of about $100,000 from use of the procedure. Waste liquor from citrus peel processing, normally a waste disposal headache, is concentrated to syrup and returned to the process. This procedure saves fuel as well as reducing the waste.

A number of industries have achieved substantial reductions in their final waste loads through by-product production. In beet sugar production, the recovery of monosodium glutamate and potash from the Steffens waste has significantly reduced the BOD generated.

In canning apples, the utilisation of the "waste" segments of the apple for vinegar production has reduced the waste load per ton. Cottage cheese whey has been converted into protein food supplements, thereby eliminating a difficult waste problem. Some of the by-products from waste materials in food processing are quite imaginative. For example, bran waste from milling has been used to make inexpensive sets of plates, casseroles, and trays, and coffee grounds have been used to produce discs and shelves.

As indicated, many alternatives are available for the reduction of wastes after generation by materials recovery and by-product production. In many cases these methods result in net profits to the industrial operation, even without counting the broader social costs avoided by the reduction or elimination of wastes discharged to watercourses.

After economic processes of materials recovery, by-product production, and effluent reuse have been exhausted, there are often residual waste loads which may have to be reduced or otherwise handled prior to final discharge. Some common methods for treating industrial wastes, also applicable to municipal wastes, are listed below:

Screening	Neutralization	Incineration
Flocculation	Chemical oxidation	Biological filtration
Chemical coagulation	Chemical reduction	Activated sludge
Flotation	Wet oxidation	Anaerobic digestion
Sedimentation	Fermentation	Stabilisation lagoons
Centrifuging	Emulsion breaking	Spray irrigation
Filtration	Evapouration	Disinfection
Stripping	Distillation	

In addition, the waste loads from some industrial operations may contain substances requiring special treatment, Such as colours, odors, special chemicals, and toxic materials.

The waste treatment methods listed above might be characterised broadly as physical, chemical, and biological processes. They have greater or lesser applicability to the two major types of wastes, nondegradable and degradable. With respect to the former, suspended and dissolved solids are the major wastes. Suspended solids can be removed by sedimentation, with or without the aid of flocculants such as polyelectrolytes, by filtration through various kinds of screens hind filters, and by centrifuging. Dissolved solids can be removed, to any desired degree, by one or more of the processes of distillation, ion exchange, electrodialysis, and reverse osmosis. It is possible to obtain completely pure water from waste discharges, but there will still be some sort of residual waste, either a concentrated brine or a semi-solid sludge. Ultimate disposal still remains a problem.

The degradable wastes of primary concern ire organic wastes, bacteria, and thermal discharges. Organic wastes are removed by physical processes, such as sedimentation, to some degree, and by chemical and biological processes, such as chemical oxidation and aerobic digestion, respectively. Bacteria are at least partially removed by biological processes and by disinfection. Thermal loads are removed by one or more types of air or water cooling systems.

Conventional treatment of domestic wastes consists basically of physical and biological processes, and is a comparatively standardised process. It can greatly reduce BOD, but will not completely, eliminate it, and is frequently designed to improve bacteriological quality and essentially eliminate suspended solids. Ordinarily the treatment of degradable organic wastes either domestic or industrial begins with the removal of the larger suspended solids by screening and grit chambers and the more finely divided suspended solids by sedimentation. New practices, such as the addition of polyelectrolytes have done much to improve the efficiency of the sedimentation process. This sedimentation stage, known as primary treatment, results in a wet, difficult-to-handle sludge, which is usually digested in heated anaerobic tanks before final disposal. In some cases the sludge is burned, usually after some degree of drying, or concentrated with centrifuges, prior to disposal. When waste treatment plants are operated adequately, primary treatment of an effluent containing an organic waste, with sludge disposal, can reduce the first stage BOD by 35 to 40 per cent.

What is generally termed a secondary stage of treatment is biological in character and essentially controls and accelerates the oxidation processes which occur in natural waters. Two major techniques are presently used, with various modifications. One is the trickling filter, in which a biological film is grown on rocks or some type of plastic medium, and the waste is applied intermittently, allowing air contact with at least the surface of the film. The other technique is the activated sludge process, in which air or oxygen is forced into a tank containing a mix of waste and actively feeding biota. Part of the settled sludge containing the biota is recirculated and mixed with the entering waste in the aeration chamber.

Each of these methods has certain advantages. Trickling filters require relatively little attention during operation, and the biological growths upon which the effectiveness of the filters depends are somewhat less susceptible to toxic substances than the activated sludge process. An activated sludge plant is somewhat more flexible in operation and design; the major variables are the proportion of sludge recirculated, the length of aeration time, and the amount of air or oxygen introduced. Rapid turnover of waste and sludge generally saves space and power, but requires well-trained personnel and continuous monitoring of plant performance. Treatment in all activated sludge plants can be made more effective by increasing the quantity of air passed through the sludge tank, although there is a definite limit to this procedure. Costs in activated sludge and other types of treatment plants increase rapidly as the removal of BOD is pushed above 95 per cent.

Small package plants, called extended aeration plants, have been used with considerable success for handling relatively small volumes of wastes, usually from strictly domestic sources. These plants are easy to operate, the sludge withdrawals are small enough in volume to be discharged with the final effluent, and much more of the solid matter is in mineralised form. When secondary treatment is undertaken, the supernatant liquor from the sludge digestor, or other liquid resulting from the handling of sludge, is routed through the secondary treatment process. Secondary treatment in turn gives rise to some additional sludge which is routed into the digestors. Primary and secondary treatment combined usually can reduce BOD by between 80 and 95 per cent. The procedures outlined above, plus disinfection of the final effluent in some instances, have been the essential elements of the primary-secondary waste treatment sequence for more than fifty years.

Although the conventional treatment processes are effective in reducing BOD, they have been much less effective in removing plant nutrients, nitrogen and phosphorus. During secondary treatment the nitrogen content of the organic waste is successively converted to nitrites and nitrates. Over-all, the volume of potential plant nutrients is reduced modestly during secondary treatment, but the content of nitrogen in nitrate form of course increases. Within limits, the extent to which nitrification occurs in the treatment process rather than in the receiving water can be controlled.

Other treatment methods that have become increasingly common and that achieve waste reduction comparable to secondary treatment are oxidation ponds (or stabilisation lagoons) and spray irrigation systems. Both methods are used in the food processing and pulp and paper industries. Ponds are used by the petroleum refining industry, by some large metropolitan areas (Melbourne, Australia, for example), and increasingly by smaller communities. Ponds, lagoons, and spray irrigation are used alone as complete treatment processes or as secondary or tertiary treatment in combination with other treatment processes. Ponds and spray irrigation systems may also be used in series, as in a number of installations in the food processing industry. Stabilisation lagoon or oxidation pond systems in industrial applications may range from a single large pond to a series of as many as eight or nine ponds. Pond capacity may be sufficient to store the entire waste effluent from a plant with a seasonal production schedule, as in some beet sugar and canning installations, or during the low streamflow period for plants with continuous production schedules. Alternatively, the pond system may be operated to produce a continuous outflow, as occurs in some installations in the canning, petroleum refining, and pulp and paper industries. A series of ponds may also have aeration devices of one kind or another between ponds and or floating or fixed aeration devices on or in one or more of the ponds themselves.

Algae play an important role in ponds and lagoons, as they furnish a substantial share of the oxygen requirements of aerobic bacteria, and they utilise the gases from the anaerobic BOD removal in the bottom sludge banks of such ponds and reduce or prevent their escape to the atmosphere. Such ponds and lagoons, when properly designed and loaded, are capable of stablizing oxygen-demanding wastes to a high degree. In the Ruhr industrial area of West Germany, where comparatively large, shallow impoundments have been constructed in several rivers to serve essentially as large oxidation ponds in the streams themselves, the impoundments are sometimes specially designed to handle the specific types of wastes they receive. One, for example, neutralizes acids and precipitates organic materials simultaneously.

One effect of the use of ponds, lagoons, and spray irrigation systems (and underground disposal) is the increase in the consumptive use of water per unit of raw product or final product, compared with conventional treatment. Data from the petroleum refining industry show that the mean consumptive use was nearly 50 per cent greater for refineries using ponds and underground disposal than for those using conventional treatment. Although the increase was small in absolute terms about 14 gallons per barrel of crude throughput it could be significant during periods of low streamflow, as in the summer months when the consumptive use in waste treatment by these methods and demands for water are both at a maximum.

It is important to note that both the oxidation pond or stabilisation lagoon and the extended iteration treatment methods differ from standard waste treatment in that their effluents contain virtually all of the plant nutrients originally in the waste discharge. In the ponds, they are embodied in algae; in the latter, in mineral form. Thus the effluents from these types of treatment processes fertilize receiving waters more than effluents from conventional treatment plants do. Considerable attention is being devoted to developing methods for removal of plant nutrients from final effluents. Some recent results suggest that progress is being made on this problem. If fish or algae were harvested from oxidation ponds, the process of biological treatment would be accelerated and a major portion of the plant nutrients, as well as some other chemicals, would he extracted from the waste water.

Chemical treatment includes a variety of processes ranging from chemical oxidation of substances in waste discharges to neutralization of acid or alkaline wastes. Most types of chemical treatment involve low capital costs and high operating costs; this makes them particularly useful for specialised treatment problems, for final finishing of a sequence of waste treatment, and for infrequent application during critical periods of water quality. Some laboratory experiments with chemical treatment have shown a reduction of 60 to 85 per cent in BOD without biological treatment. Disinfection, by use of chlorine or ozone, can also be considered a type of chemical treatment.

One final point should be made concerning waste treatment. The increasingly stringent quality standards of final effluents make the economics of in-plant water recirculation more favourable. In some cases, water may be suitable for reuse within a plant even as process water and yet not meet the quality standards for effluents. The other major type of degradable wastes consists of thermal loads, which are of increasing importance. The most common method of reducing the heat load discharged is the installation of cooling towers. The magnitude of heat load reduction possible by this process is exemplified by the installation at the Philadelphia refinery of the Atlantic Refining Company, where the cooling tower program resulted in an 85 per cent reduction of heat load discharged to the river.

Whether or not a cooling tower is installed at a particular plant is a function of both the internal economics of over-all water utilisation in the plant and of the offsite costs resulting from the discharge of thermal loads. With respect to cooling towers, the economics of recirculation are such that if recirculation is adopted, it will be done essentially completely. As a result, little or no heat will be discharged, but there will be an increase in the discharge of dissolved solids, stemming from chemical treatment of the recirculated water and concentration from evapouration.

Cooling towers may also aid in reducing other waste loads. For example, in the Toledo refinery of Sun Oil Company, a cooling water stream containing phenols passes through a cooling tower which functions essentially as a trickling filter. Biological oxidation reduces the phenol content in the effluent by over 99 per cent. Other applications of the same procedure have been reported.

A final method for handling residual wastes is the reuse of a municipal or industrial effluent by a subsequent user or users, usually after one or more types of man-made or natural treatment. Sewage effluent is used for various purposes in industrial operations, particularly in the southwestern United States. The practice is less common in the East, although effluent from the Baltimore municipal treatment plant is used in the Sparrows Point plant of Bethlehem Steel, primarily for cooling water in steel production. Effluents from municipal waste treatment plants have long been used for irrigation.

Among the many cities that use all or part of their sewage effluents for irrigation are Grand Canyon and Tucson in Arizona; Bakersfield, Fresno, San Bernardino, and San Francisco in California; and over 200 towns in Texas including Abilene, Kingsville, Lubbock, Midlands, San Angelo, and San Antonio.

Different reuse methods have varying effects on residual waste loads. Irrigation tends to remove virtually all of the residual wastes from the effluent, while industrial reuse may or may not reduce the residual waste load. Irrigation use, however, results in considerable net depletion of water, which is not necessarily unproductive. Artificial recharge of groundwater aquifers represents another method of effluent reuse. In this case, however, generally a larger number of intermediate processes are involved than in the cases mentioned previously.

Usually the effluent from a municipal waste treatment plant is inserted into a groundwater basin by means of injection wells or spreading grounds. In the latter case, the wastewater undergoes additional purification as it passes through the unsaturated zone. The purification process in the soil results from the action of soil bacteria, filtration, adsorption, and perhaps other processes. In other cases the waste effluent from the sewer system is conveyed directly to a wastewater reclamation plant, as at Whittier Narrows, California, where it is treated and then used for groundwater recharge. Another extensive system has been in operation for several years at Santee, California, where treated effluent is discharged into a series of ponds, the final ones having now been approved for water-contact sports. The final outflow is recharged to the groundwater basin. Caution is necessary in using wastewater from domestic and/or industrial operations for groundwater recharge. Even with treatment the minerals in the effluent before recharge remain, so that the total dissolved solids content is likely to increase (although there may be wide dispersion in the groundwater aquifer). Further, because there is likely to be some residual load even after treatment and the process of groundwater recharge, it is possible for groundwater contamination to occur. However, the well-publicised quality degradation of groundwater resources in certain areas, such as Long Island and Minneapolis-St. Paul, happened in situations where there was no centrally designed and operated wastewater treatment and recharge facility. Recent analysis suggests that groundwater recharge for water renovation may have much wider applicability and be more economical than has previously been appreciated.

An important point in connection with all measures of handling wastes is the existence of economies of scale. In materials recovery this is illustrated by the costs of the Blaw-Knox Ruthner sulfuric acid recovery process for handling waste pickle liquor. On the basis of 8.5 per cent free acid in the liquor, costs are about 3.5 cents per gallon when 30,000 gallons per day are processed and about 3 mills per gallon when 160,000 gallons per day are processed.

Similarly, significant economies of scale exist in waste treatment. For example, in a study of fruit and vegetable canneries, the costs (annual charge on capital investment plus operation and maintenance costs) for lagoons and/ or spray irrigation systems were found to vary from about 2 cents per pound of BOD removed for operations removing more than 500,000 pounds of BOD in a canning season to 30 cents per pound for operations removing less than 100,000 pounds. Extensive analyses by the U.S. Public Health Service show similar economies of scale for municipal waste treatment plants. In densely developed areas the costs of collective facilities for conveying and treating municipal and industrial wastes can often be much less than the sum of the costs of separate facilities for each municipality and industrial plant.

Reducing Waste Discharges

The foregoing discussion has focused on the wide variety of methods by which the quality of receiving waters can be improved by reducing waste loads discharged to them. The various methods can be combined in a myriad of ways to meet the needs of a single industrial operation or urban area or a complex consisting of many operations and or areas. Considcer just waste treatment to reduce waste loads. Even in a single industry the combination of treatment processes applied in different plants varies widely, primarily as it function of the raw materials utilised, the production processes involved, the final products, and environmental constraints, including effluent controls. Consequently, there is a wide range in waste treatment costs in any given industry.

What can be done to reduce waste loads discharged to receiving waters is clearly exemplified by the performance of the beet sugar industry. The types of measures to reduce waste loads used by the beet sugar industry are likely to be integral components of optimal water quality management systems. The planning of such systems requires the development of relationships between the incremental quantity of wastes removed and the incremental cost of removal. .

Increasing the Assimilative Capacity of Receiving Waters

Because of the inverse relationship between the concentration of most water quality variables and water quantity, providing water for dilution by fuller use of available streamflow through flow regulation suggests itself as a possibility. Low streamflows often coincide with heavy concentrations of waste loads, as well as with high temperatures. Where the waste loads consist of oxygen-demanding wastes, the dissolved oxygen level of receiving waters is thereby depressed. Periods of low flow and potentially large reductions in water quality are the conditions for which effluent treatment plants and effluent redistribution systems are generally planned and designed. The assimilative capacity of tidal estuaries may also be significantly affected by variations in inflow; but the effects are more dispersed and difficult to analyse than in the case of flowing streams. Dilution of wastes beyond that normally provided by natural streamflow during low-flow periods can be provided by modifying the pattern of streamflow, by modifying the time pattern and 'or locations of waste discharge, or by some combination of the two. The most common practices are to increase flows during low-flow periods by controlled releases from reservoir storage, and to modify the pattern of waste discharge by temporary storage of wastes. Flow can also be increased by withdrawing water from groundwater sources and releasing it into surface watercourses. An interesting variation of low-flow augmentation is the recently inaugurated Buffalo River Improvement Project, in which water is pumped from Lake Erie, used as cooling water on a oncethrough basis in five industrial operations, then discharged into the Buffalo River to maintain a minimum flow.

The effectiveness of augmenting flow depends on both the type of waste and the type of receiving water involved. With respect to organic wastes, for example, increasing assimilative capacity by flow augmentation is effective for streams, but not for lakes, and only to it limited extent for estuaries. Increased flows could actually reduce water quality in lakes and estuaries by carrying greater quantities of partially assimilated wastes into them. With respect to nutrients, increased flows would likely result only in a more rapid discharge of these substances into lakes and estuaries, because less time would be available for adsorption and sedimentation in stream channels. In contrast, with respect to nondegradable wastes such as chlorides, increasing, the flow into estuaries can result in a significant improvement in water quality. It appears difficult, if not impossible, to increase the assimilative capacity of groundwater bodies by augmenting inflows.

Flow augmentation may play a large role in future water resources management in the United States. Large-scale planning efforts of the U.S. Corps of Engineers in several basins in the East will include flow augmentation for water quality improvement its an important feature. The final report of the Corps on the Potomac River Basin recommended a large amount of reservoir storage for increasing low flows to improve water quality.

The effects of reservoir storage itself on the quality of water later released may be favourable or unfavourable. On the positive side, bacteriological quality tends to be stabilised, and summertime releases from reservoirs tend to be cooler than normal streamflows. On the other hand, some early hopes for effective dilution through releases from reservoirs were disappointed because of the failure to consider the impact of impoundment on the dissolved oxygen content of releases. Water from deeper parts of reservoirs is often virtually devoid of oxygen, because of the combined effect of BOD demand and reservoir stratification. When this condition is present as it is especially likely to be in areas experiencing hot summer seasons it takes larger releases from reservoirs to achieve it given effect on dissolved oxygen downstream, and the costs of achieving the desired water quality level are increased. There are several ways of coping with the unfavourable effects of water storage in reservoirs. Multiple outlets can be installed as at the Oroville Dam storage unit on the Feather River in California so that water can be released from different levels in various combinations to achieve the desired water quality. If there is a power installation at a reservoir, air or oxygen can be introduced into water in the turbines. In many cases this can be done rather simply by using vacuum breakers already installed in the dam. Turbine aeration involves some power loss, and therefore is not costless.

Reservoirs can have significant effects on other elements of water quality such as suspended sediment, turbidity, and certain chemical constituents. An in-stream reservoir acts as a settling basin, and will eliminate or reduce problems of suspended sediment or turbidity. The clearer water may mean lower treatment costs for downstream water users, or it may promote increased algae growth with associated taste and odor difficulties. Another method for modifying water quality in impoundments (and in natural lakes) is by mixing. The purpose is to destratify the impoundment or to prevent stratification from occurring. A number of installations have been made in the last ten years. Results show that it is possible to destratify an impoundment by mixing and to improve the quality of the entire water mass, but they also show that stratification can reoccur after mixing ceases. Hence, prevention of stratification appears to be the better approach. The desired circulation patterns can be achieved through reservoir designits shape and location of inlets and outlets as well as by mechanical mixing.

The degree to which the assimilative capacity downstream from an impoundment is improved is dependent not only on the reactions which take place in the Impoundment but also on how the reservoir is operated. Depending on the time pattern of releases, there can be favourable or unfavourable effects on the assimilative capacity. For example, when a power plant at a reservoir is used for "peaking," the release pattern will vary substantially. Flows will be high during the portion of each weekday when the peaking load is being met, and low during the night and on weekends. This pattern is in contrast to the more continuous pattern of waste discharges. Reregulation of the power releases can compensate for the varying flow, but of course at increased cost.

Another way of increasing the assimilative capacity of streams and lakes is by artificial reaeration with either air or oxygen. Mobile or fixed aerating devices can be installed in reaches of streams where needed to replenish the dissolved oxygen supply and prevent anaerobic conditions. Reaeration differs from methods for modifying wastes prior to discharge in that nothing is removed from the receiving water, and from methods for increasing the degree of dilution in that there is no increase in streamflow velocity to alter the length and shape of the oxygen sag. The effect is analogous to that of a lengthy oxidation pond artificially aerated. In various coastal areas of the United States the heavy use of groundwater resources has led to saltwater intrusion. One solution to this problem is to reduce the pumpage from the affected groundwater aquifers. Another is to establish barriers to the intrusion. In Southern California, the Manhattan Beach Project for reducing seawater intrusion and preventing further deterioration of the available groundwater supply created a saltwater barrier by a series of injection wells. The well system provides a groundwater gradient toward the ocean which prevents further intrusion and enables the continued use of the groundwater aquifer. Impermeable physical barriers are another way of preventing saltwater intrusion.

Surface water is also subject to saltwater intrusion. Decreased outflow from upstream areas of rivers which empty into tidal estuaries has resulted in sporadic seawater intrusion, as in San Francisco Bay and the Delaware estuary. This type of intrusion can be reduced by increasing the flow into the estuary, or by constructing physical barriers to upstream saltwater movement. The first method has been utilised for many years in connection with San Francisco Bay, where the Central Valley Project has been operated to provide releases to reduce seawater intrusion. The second method has been proposed and studied for both San Francisco Bay and the Delaware estuary. The difficulty with physical barriers is that they impede navigation, affect aquatic life, and aggravate problems of managing wastes discharged upstream. Seawater encroachment in tidal estuaries will become more serious, as upstream development increases the consumptive use of water. Effluent redistribution to make more efficient use of the available assimilative capacity can be achieved in a number of ways including regulated discharge, transfer of the effluent to a different location for discharge, and underground disposal. In general, these methods have little, if any, effect on waste loads as such, and they do not modify the receiving water.

Regulated discharge involves the temporary storage of waste effluents for discharge at times when more water is available for dilution. In an estuary, this may mean storage during the ebb tide and discharge on the flood tide. Often it involves storage of wastes during low-flow periods in the summer and discharge during winter and or spring when streamflow is higher. Some degradation of wastes takes place during seasonal storage, and this procedure is used by a number of foodprocessing, pulp and paper plants, and petroleum refineries. The efficient use of regulated discharge depends on sophisticated studies of the operation of such a system, and on a data collection system that provides continuous information on the quantity and quality of receiving water and the quantity and quality of wastes to be discharged. Transferring of wastes from one location of potential discharge to one or more other areas where the assimilative capacity is larger is usually practiced in areas near tidal estuaries or oceans. This system is now used by the Toms River Chemical Corporation, which formerly handled the wastes generated in the production of synthetic and organic dyes by equalization, neutralization, clarification, chlorination, aeration, and discharge to the adjacent Toms River. When the residual waste load from growth in production approached the assimilative capacity of the river, ocean discharge was adopted. This made possible the elimination of both chlorination and aeration. The entire waste-handling system is continuously monitored.

In underground disposal, effluents are discharged into relatively deep formations that contain so little water or water of such poor quality that they are not expected to be used for water supply. This method of disposal is used largely by the chemical, petrochemical, and paper industries for wastes that are difficult or very expensive to treat. Underground disposal represents a consumptive use of water, since the effluent is not available for reuse.

The wide variety of measures for water quality management discussed above could be incorporated in plans for a water utilisation system in a region (say a river basin). Planning, however, is only the first step in the management problem. The system must be implemented and once in existence must be operated to produce a specified time pattern of outputs of water and water-related goods and services, including water quality, at specified locations. This it must do in the face of uncertainties relating to hydrologic events and to demands on the water resource over time. Efficient and effective operation of a water utilisation system involves not only day-to-day operation of individual units in the system, such as waste treatment plants and reiteration devices, but integrated operation of all units in the entire system. This is no mean task. Before concluding our comments on the technical and engineering aspects of water quality, we turn to a short discussion of system operation.

Operating a system means determining when to push buttons, turn valves, open and close gates, and so on for the various units involved in, and economic activities related to, it water utilisation system. Such operation requires a communications-control network consisting of data collection, data communication, data analysis and interpretation, signaling, and response, his illustrated in the generalised collection involves continuous recording of streamflow, precipitation, soil moisture, continuous monitoring of various water quality parameters in receiving waters and waste discharges, continuous recording of the volume and quality of water in storage reservoirs and in groundwater basins. In the last decade, considerable advances have been made in the development of instrumentation for continuous analysis of various water quality variables such his dissolved oxygen, temperature, turbidity, pH, total dissolved solids, conductance, and COD (chemical oxygen demand). From the data collection points the information must be transmitted to a control centre. Again, recent developments improved this component of the management system through such techniques is automatic telemetering of data. In the analysis and control centre, the data are compiled and printed out or stored. Assuming basic relationships have already been developed, such as among streamflow, waste loads, and water quality, and streamflow and stream velocity, estimates can be made of water quality at various points in the system. This information is relevant to both shortrun and long-run situations. Examples of the former are accidental spills of toxic materials and intense storms producing high concentrations of suspended sediment which might clog water intakes or groundwater recharge facilities. The operators and water users at various locations can then respond to the "signals" received in terms of pushing buttons, turning valves, opening and closing gates. Artificial reaeration devices can be started; releases from reservoirs can be made; additional chemicals added in various treatment processes. Artificial groundwater recharge facilities may be by-passed. If necessary, generation of wastes can be reduced by curtailing production.

Despite all of the problems relating to system operation, the net gains which can he expected from the development of what we have called a communications-control network appear to be significant. Such networks were probably conceived first to preclude serious damage from accidental spills of toxic materials or substances that would disrupt waste treatment and 'or water treatment facilities (e.g., phenols, cyanide, arsenic) or to monitor the performance of waste treatment plants in relation to receiving waters. As demands on the water resource increase, with a concomitant increase iii interactions among various water users, the need for, and the potential savings from, sophisticated system operation will increase. Savings stem from the possibility for more nearly optimal system plans, i.e., reduction in size of some system components, and from the possibility of incremental operation of water quality management systems once they are in existence. The latter refers to the use at any given time of those measures which provide the desired water quality at least cost. The benefits from this type of system operation, including computer control, are well documented in the power industry.

It should be emphasized that the planning of any water management system involves assumptions about the way the system components are to be operated when they are finally In place. Accurate planning *requires* determination of the performance of proposed systems under various possible hydrologic events. Accordingly, to achieve a valid assessment of system performance requires structuring the simulation of the water utilisation system in the planning phase in a manner as close as possible to the situation that will exist in the real world. Even so, the efficiency of operation of system components assumed in the planning phase can be approached, but never achieved, in actual system operation. The degree of articulation and integration of water quality management system components implied in the above paragraphs has so far not been achieved even in system operation, much less in the planning of actual systems. The Ohio River Valley Water Sanitation Commission, however, has made substantial progress on the operation phase, especially with respect to a communications network. In most regions, water quality management is at a primitive level. At the same time, the technical and analytical means for implementing such systems are developing rapidly. Institutional means lag far behind. Our discussion of the engineering and technological facets of water quality management has simplified many of the processes and relationships, but one central point should be clear. Technologically, there are many alternative measures available to achieve whatever levels of water quality ire desired. Those include various types and degrees of intake water treatment, measures for reducing the generation of wastes, measures for modifying and disposing of wastes after generation, and measures for increasing or making better use of the assimilative capacity of receiving waters.

These measures substitute for one another economically at various rates depending upon the particular waste loads and receiving water bodies involved. Decisions concerning the proper balance among the various measures cannot be made rationally on technical grounds alone, because not all of the benefits from water quality improvement can be measured in dollar terms. Accordingly, values must be introduced into the decision-making process. Further complicating the decision process are both the stochastic nature (uncertainty of occurrence) of hydrologic phenomena and the inadequate understanding (as yet) of the impacts of waste discharges on water quality of various receiving waters and of the effect of various levels and time patterns of water quality on various water uses.

Identifying aternatives for a water quality management system and devising plans which combine them optimally is a demanding task. Moreover, planning and implementation are complicated by the fractionation of decision-making responsibility for the various possible components of such systems. Some measures are within the purview of the individual water users and waste dischargers, some are the responsibility of state, regional, or federal agencies, and some do not fall within the conventional purview of any agency whatsoever.

The problem is further compounded by the fact that water quality improvement is generally only one of the desired outputs from a water management system. Both system planning and operation are affected by the complementary and competitive relationships among the various outputs water quality improvement, energy generation, navigation, flood damage reduction, irrigation, water supply, and water-based recreation. We now turn to the economic aspects of regional water quality management.

6

The Economic Value of Improved Wastewater Irrigation

Wastewater is increasingly being used for irrigation in urban and peri-urban areas as well as in distant downstream rural areas of large cities in developing countries. The practice of wastewater use in crop production is also common in Ethiopia, the second most populous country in Sub-Saharan Africa. It is essential for the livelihoods of many poor farm households, and supplies fresh vegetables at low prices to the nearby cities. Moreover, it serves as a means of supplementary income for small businesses through the vegetable marketing channels. Without adequate safeguards, however, wastewater irrigation can cause serious drawbacks to public health and the environment. These positive and negative consequences of wastewater use challenge decision makers to identify practical and affordable strategies for the safe use of wastewater that do not threaten the various livelihoods depending upon it. Addressing this challenge requires identifying, assessing, and evaluating the negative and positive impacts of wastewater use in crop production, based on a comprehensive analysis of economic costs and benefits.

The use of wastewater for irrigation is associated with adverse effects on farmers, public health, and the environment. Farmers are affected by direct contact with contaminated wastewater, and its use in agriculture causes negative externalities both to public health through the consumption of agricultural produce irrigated with wastewater and to the environment by constantly polluting the groundwater and soils. In Ethiopia, both domestic and industrial wastewater is discharged mostly untreated into the nearby rivers, which are used as sources of irrigation water. Hence, smallholder farmers cannot enforce their right to clean water, and have no control over the availability and quality level of the wastewater used for irrigation. Since farmers in and around the cities depend on the use of wastewater for irrigation, adequate policies are needed to ensure farmers can use it safely for crop production without negative externalities to the health of consumers and the environment.

Figure: *Recycling wastewater for irrigation. Wastewater in process*

The main objective of this study is therefore to assess the value farmers attach to safe use of wastewater as a basis for devising government policies for regulating wastewater discharge, treatment, and agricultural use. Since wastewater is a nonmarket good, this study employs the contingent valuation method (CVM), which is regularly applied in the field of environmental valuation. Farmers' willingness to pay (WTP) for policies to ensure such a safe use of wastewater for irrigation is used here as an indicator of the welfare gain of those policies. Specifically, we assess two different policy options, as well as a combination of the two, in order to determine their suitability for the assessment of the value of safe wastewater use. Moreover, the study also compares the value of safe wastewater irrigation with the reference case of freshwater irrigation for crop production and assesses comparable policy options for that case.

Though the CVM is in principle applicable in developing countries, only a few published studies have applied this method to forests and environmental protection in Ethiopia. To our knowledge, there has been no study published using the CVM to compute the value of irrigation (waste) water in this country. We believe our study is the first of its kind in using this technique for valuing wastewater resources in Ethiopia. The remainder of the chapter is organised as follows. The next section describes the methodological approach of this study and lays out the challenges to applying the CVM in the context of Ethiopia, and briefly describes the study area of our survey. The third section presents the survey design of the actual CVM survey carried out in and around

Addis Ababa. This is followed by a section that lays out the empirical model used for data analysis. We then discuss the results from the descriptive and econometric analysis, and estimate the welfare gain from safe use of wastewater for irrigation. The final section presents a summary and conclusions.

Assessment of Welfare Changes Using the CVM

A typical CVM survey entails three operations that offer many possibilities for methodological variation: (a) designing a survey questionnaire, (b) conducting the survey, and (c) analysing the survey results using econometric techniques. Although there is no standard approach to designing a contingent valuation questionnaire, it should contain the following basic components: a detailed description of the project scenario(s) respondents are asked to value, a description of the payment mechanism for the public good and of the conditions under which it will be provided, the WTP elicitation question, and finally, some debriefing questions regarding respondents' attitudes toward the project and their socioeconomic characteristics. From the responses, one can estimate the average WTP per household, which, by convention, is multiplied by the number of households affected by the project to obtain the overall change in the population's welfare resulting from the project.

Although the CVM has become a standard procedure for the assessment of economic benefits resulting from nonmarket goods in developed and industrialised countries, its application in developing countries such as Ethiopia poses a number of challenges. In this study we assess whether the CVM can be employed in Emiopia as a tool for modern cost-benefit analysis in the context of a rather complex public project. Apart from the requirement that respondents have sufficient cognitive capabilities to understand the public project proposed to them and to envision the benefits it would entail for them, the greatest challenge in developing countries is the widespread lack of trust in government and the authorities. Respondents need to perceive that the proposed project will actually lead to the promised improvements and that their responses given during the interviews will influence the chances of the project's implementation. Also, respondents must be certain that their individual financial contributions will be used exclusively for the project. Another serious limitation of the CVM is the often very low level of disposable income, especially of smallholder farmers, which might cause them to restrict their stated WTP based on their low ability to pay. In such a situation, welfare gains would be systematically underestimated. In this study, we address these issues to scrutinize the applicability of the CVM in the Ethiopian context.

The Empirical CVM Survey

This study was conducted within and around the city of Addis Ababa. The population of the city is currently estimated to be about 5 million, with an annual growth rate of 2.9% (Ethiopian Economic Association, 2005). Environmental sanitation services are very limited in the city. For example, discussions with the city's water and sanitation officials reveal that less than 15% of the population has access to adequate sanitation services and less than 1% of households are connected to sewer lines. Moreover, 35% of the solid waste is disposed of improperly. The existing publicly owned treatment plants treat no more than 1% of the total annual domestic wastewater production of 3 1 million cubic metres. Industries in the city, constituting about 65% of the country's total, do not have functional wastewater treatment or disposal systems connected with the city's network. These industries produce about 5 million cubic metres of wastewater per year. Almost all the wastewater generated from the city is discharged directly into the rivers or streams flowing within the city without any kind of treatment.

These rivers and streams are tributaries of the Akaki River, which is the source of irrigation water for vegetable growers. Since it serves as a sink to the wastes from the city, the river is highly polluted with organic and inorganic substances, and the pollution level, as measured by biochemical oxygen demand (BOD), phosphates, suspended solids (SS), ammonia, and total coliform bacteria, exceeds by a large margin the standard values set by the Ethiopian Environmental Protection Authority. The pollution level varies among the different streams of the river and between the different seasons, being higher in the downstream areas and during dry and short rain seasons. Farmers are using water from the Akaki River to produce vegetables both for market and for home consumption. The crops grown are the main source of income for the villagers, so their livelihood is directly linked to the untreated wastewater. The share of wastewater farm income ranges up to 97% of total annual household income. The wastewater farms also provide about 61% of the vegetables supplied to Addis Ababa, and many poor traders depend on the vegetable market for their survival. Clearly, the unsafe use of this wastewater for crop production raises specific potential health concerns. The crops grown include leafy and root vegetables, which can be eaten raw, and the irrigation techniques can expose farmers to the contaminated soil and wastewater, especially since they do not wear protective clothing during farming. In addition to the health risk to farmers and consumers, the wastewater also adversely affects the environment by polluting the air and contaminating the soil. Overall, even though the wastewater from the city is an appreciated resource, its use in agriculture must be carefully managed in order to maximise the substantial benefits at minimum health and environmental risks.

A comparison group of farmers using freshwater for irrigation are located about 40 kilometres east of the centre of the city. Similar to the wastewater group, these farmers reside within the Awash River basin. Both groups share similar agro-ecological and climatic conditions, produce similar irrigated crops, and are full-time farmers. The freshwater group is in a rural area where drinking water supply and other basic infrastructure are at a lower level compared to their wastewater counterparts. Problems occurring in that area linked to irrigation include health concerns (water-logging can create favourable circumstances for mosquitoes, and thus for malaria) and sustaining productivity of the soil, because improper irrigation practices can lead to erosion.

Data Source and Sampling

This study uses mainly primary cross-sectional data for the year 2006. The main data source is a CVM survey conducted on both wastewater and freshwater farm households within and around Addis Ababa city. A double-bounded dichotomous- or discrete-choice format was used to elicit respondents' WTP for improved or safe use of wastewater for crop production. The survey areas were selected because the farmers are performing irrigated farming activities using irrigation water of varying quality. This enables us to estimate the value of the wastewater and compare it with that of freshwater in irrigation.

To ensure homogeneity in the grouping of the households, the wastewater farmers were divided into four subgroups depending on the location of the farm areas. These include the Kolfe, Lafto, Akaki-Addis, and Akaki-Oromiya farm areas. We also divided the freshwater farmers into two subgroups depending on the source of irrigation water. These are the Godino farm area, where farmers use diverted water from the Godino River, and the Fultino farm area, where farmers use spring water for irrigation.

The distribution of the number of samples between the freshwater and wastewater areas was determined based on the number and proportion of farm households engaged in irrigated agricultural activities within the respective areas, and the available financial resources. Accordingly, a total of 415 sample farm households were included in our survey, of which 175 were freshwater and 240 were wastewater farm households.

Before the main survey, the enumerators were given one day of training, mainly focusing on the technicalities of the questionnaire, and a pilot survey was conducted to check its wording, ordering, and timing, and to determine the starting bids for the dichotomous-choice valuation question in both wastewater and freshwater areas. In the pilot survey, an open-ended elicitation format was employed to determine the starting bids. The maximum WTP obtained from the pilot survey ranged from zero to ETB 100 per year per hectare. We used the most frequently occurring WTP values of ETB 20 and ETB 40 as starting prices to elicit the value to households of safe wastewater use/proper use of irrigation water for crop production. Respondents were divided randomly between these two starting prices. The survey was administered as in-person interviews. During the interviews, efforts were made to collect information from the head of the household. In the event the household head was unavailable, the interview was conducted with a member of the household above the age of 18, who was able to provide reliable information about the household.

In our CV survey, respondents were asked about their current use of irrigation water (wastewater or freshwater), the benefits of using irrigation water, the potential risks from unsafe use of irrigation water, and the advantages of improved (safe) use of irrigation water. Subsequently, government policies were described to respondents which aimed at improving the current system of irrigation water use. These policies would result in both health improvements of farmers and the general population, as well as in preserving soil fertility and quality. Further, the circumstances under which these policies would come into effect were explained to respondents.

Wastewater farmers were given the choice between two government policy options. The first policy option would focus on advantages to the farmers by introducing a safe wastewater irrigation practice. This would be achieved by implementing a training program offered by the local department of agriculture that would increase the problem awareness regarding wastewater irrigation among farmers and would introduce and implement simple, available, and affordable measures to be taken during farm work, harvesting, and consumption of the produce. The second policy option would address the polluters' side by implementing a government program to rigorously enforce existing anti-pollution laws.

The concentration of contaminants in effluents and surface water bodies would regularly be monitored and unlawful emissions of pollutants by households and firms would be legally prosecuted by the environmental protection authority. Consequently, improved or safe use of wastewater for irrigation was described as the possibility of using wastewater for irrigation with little or no risk to the health of farmers and consumers so that farmers could continue to use the wastewater without any concern that the municipality would eventually ban this practice. For respondents who were using freshwater for irrigation, the policy option presented pursued similar general objectives; however, they were adapted to the specific situation of freshwater farmers. The policy program (including technical elements and advice to the farmers) would minimise health risks to the freshwater farmers and avoid soil erosion, thereby guaranteeing long-term sustainable yields.

Following the scenario description, the payment mechanism was introduced and the WTP question was asked. The payment vehicle was implemented in the form of an annual water charge per hectare to be paid just after the harvest, which would contribute to the costs of the described governmental programs. Respondents were told that in the event the program were implemented, they would actually have to pay such a water charge.

The WTP question was asked in a double-bounded dichotomous-choice format with an open-ended follow-up. In the first question, a respondent was asked whether she was willing to pay a certain specified amount of money (either ETB 20 or 40). If she responded "yes" to the initial bid, the respondent would be asked the same question but with a higher amount of money (either ETB 30 or 50). If the respondent's response was "no" to the initial bid, a follow-up question with a lower amount was asked (either ETB 15 or 30). This was then followed by an open-ended question in which each respondent was asked to express her maximum WTP for the chosen scenario. Just before asking the WTP question, respondents were told to consider their available budget and were reminded that the annual water charge would decrease their incomes.

Finally, in designing and conducting the questionnaire, attempts were made to minimise biases such as strategic, hypothetical, and compliance biases, which may arise in a CVM survey. In attempting to minimise strategic biases, we explained the purpose of the study explicitly so that the respondent would not strategically respond to influence the provision of the program. This was done both at the beginning of the survey during focus group discussion, and during the main survey by explaining that the primary purpose of the survey is academic. In order to reduce compliance bias, we carefully developed the questionnaire, conducted a pilot survey, and trained the enumerators, who were quite experienced in household and contingent valuation surveys. In addition to the respondents' WTP, a range of information was collected in our CV survey, including the socioeconomic characteristics of the farm households, the farming system, irrigation water management, asset ownership, income, and expenditures at the household level.

However, a prerequisite for using this empirical model for double-bounded responses, also called the "interval-data model", is that respondents' preferences regarding the proposed public good remain the same when answering to the first and the second bids. Technically, the error terms of the responses to the first and the second bids are assumed to be perfectly correlated, i.e., the correlation $<1 = 1$. While this is a plausible and desirable assumption from a theoretical point of view, it is not a trivial one. It has been shown repeatedly in the literature that the second answer may be influenced by the answer to the first bid, where this first bid serves as an anchor. In such a situation, it would be more realistic, and thus appropriate, to model the double-bounded responses by means of a bivariate probit model which allows for <1 being less than one. The response probabilities to the first and second bids then can be expressed by making use of the bivariate standard normal distribution $98(B_{i1}, B_{i2}, 8_1, 8_2, <1)$, with B_{i1} and B_{i2} being the first and second bid, respectively, and <1 denoting the correlation between the error terms. The higher the correlations between the two error terms, the more consistent are the responses to the consecutive bids, i.e., the less respondents anchor their response to the second bid on their response to the first. Consequently, the interval-data model is a special case of the more general bivariate model with $8_1 = 8_2$, equal variances, and <1 = 1.

In a Monte-Carlo simulation study, Alberini (1995) computes welfare estimates using an interval-data and a bivariate probit model in comparison and finds that the results of the interval-data model are rather robust for values of <1 around 0.9 (indicating imperfect correlation between the responses to the first and second bids) in the case when covariates are used. In models without covariates, the interval-data model is still astonishingly robust even with values of <1 being much lower. However, since the interval-data model yields smaller variances of the parameter estimates, and hence is statistically more efficient, it is suggested to apply both model specifications in parallel and select the results from the interval-data model if <1 has been found to be sufficiently large. We follow this suggestion here, and it will constitute an important criterion for judging the validity of our empirical results.

Basic Characteristics of the Existing Irrigation Systems

Among all households interviewed, 14% are female-headed and the remainder are male headed households. In the study area, the average household head is 47.6 years old if female and 42.2 years old if male. Of all respondents, 72%o are heads of the household, and the rest are other household members above the age of 18 years. The education levels of the household heads range from unable to read and write to college graduate.

About 77%o received at most four years of formal schooling, while approximately 13% attended second-cycle primary school (5-8 years of formal schooling) and 10% attended at least nine years of formal schooling. About 75% of all the respondents are married, and the average family size in the study area is 5.3 persons. The average family size of a wastewater farm household is 5.6, of whom 3.6 members are of productive age (15-64 years of age); the remainder arc dependents (below 15 and above 65 years old). Freshwater farm households have 4.9 members, of whom 2.9 are of productive age. Irrigated agricultural activity is the main source of income for the farm households. The average irrigator has 0.72 hectares of irrigated farmland, which varies between the two groups of farmers (0.4 hectares for the average wastewater farmer and 1.16 hectares for the freshwater farmer). Wastewater farm income accounts for 62% of annual total household income, ranging from 27% to 97%. A freshwater irrigator receives about 67.6% of his or her total annual income from irrigated agriculture. This highlights the importance of irrigated farming to the livelihood of the farmers. Other sources of income, in order of importance, are off-farm income, livestock income, and remittances. Farmers use irrigation water for multiple purposes. Of the total sample, at least 70% of the farm households use irrigation water for domestic use in addition to crop production. Comparing the two farm groups, nearly 49% of the wastewater farmers and all of the freshwater farmers use irrigation water for both growing crops and domestic purposes. Domestic uses of wastewater include bathing, laundry, livestock drinking, and washing of vehicles, bags, and other items. However, this use varies among the different locations within the wastewater farm areas. The upstream and middle-stream farmers use the wastewater only for irrigation purposes, whereas the downstream farmers use it for all the purposes identified above.

Our study shows that even if irrigation water poses potential health risks to both farmers and consumers, farmers perceive that its benefits outweigh the health hazards. Of the total sample, nearly 85% state that they benefited a lot from the use of irrigation water. About 88% of the wastewater farmers report that it is their only means of survival. Income from wastewater farming was the largest share in the total annual income of the farm households. Only 3% of the wastewater farmers indicate that they benefited only a little, though they still work on the farm for lack of livelihood alternatives. On the other hand, about 21% of the wastewater farmers and 31% of the freshwater farmers perceive that their health is affected by the irrigation water. This difference is mainly because the freshwater farmers are affected by malaria due to the higher prevalence of mosquitoes in their surroundings.

The wastewater farmers also accept the notion that working on wastewater farm areas and consumption of vegetables produced using the wastewater has health risks, which is as expected. This perception differs between the downstream and the middle (Lafto area) and upstream (Kolfe) farm areas because of variation in the pollution level, which is higher in the downstream areas of the Akaki River. As anticipated, the perception of risk is stronger in downstream areas than in the middle and upstream areas. The wastewater causes a bad odour in the surrounding area, and this effect is worse in the downstream areas as compared to the middle and upstream areas due to the difference in pollution levels. Obviously, the current use of wastewater for vegetable production is unsafe for the health of both farmers and consumers. Wastewater farmers are more likely than freshwater farmers to suffer from parasitic infection diseases.

The wastewater is also polluted with chemicals and bacteria that can cause diseases for farmers. The vegetables produced by wastewater contain metals that potentially cause illness for consumers. Because farmers use furrow irrigation methods with bare feet and carry out weeding and harvesting with their bare hands, they are exposed to the wastewater and the contaminated soils. Moreover, almost all farmers and their family members consume raw vegetables directly from their fields without any treatment. This situation needs to be improved so that farmers can use the wastewater for vegetable production safely or with minimal health risks to their families and to consumers.

Improving the Existing Irrigation Systems

As described above, wastewater farmers in the survey were presented with different policy options to achieve the goal of safe use of wastewater irrigation from which they could choose their preferred option (either one of the two, or both). This procedure was used to ensure respondents only value those options that they deem credible and feasible. Among the different options presented, the second option, "forcing emitters to comply with the existing environmental regulations," is the most preferred.

Nearly 40% of the wastewater farmers voted for implementing only this option. About 21% of the wastewater farmers, however, preferred "awareness promotion on the methods of safe use of wastewater" as an option to improve the situation. This includes awareness about improved irrigation management techniques, types of crops grown, safe consumption of raw vegetables, quality protection during harvesting and marketing, and provision of protective dress at affordable prices. Finally, about 38% of the wastewater farmers prefer that both options be implemented jointly (i.e., regulation enforcement and awareness promotion).

Our survey reveals that 98% of freshwater irrigators are willing to contribute to improve the existing irrigation practice to minimise the malaria risk and soil erosion, while 90% of wastewater farmers are willing to contribute to a program designed to improve the existing situation aiming at safe wastewater irrigation. Only 7.5% of the wastewater farmers responded negatively to the improvement program. Three main reasons are provided by this latter group. First, the farmers state that they are paying land tax to the government, and therefore the government should protect the environment. Hence, this group of respondents might be classified as protest respondents. Second, since they are already using the wastewater properly and have not experienced any noticeable adverse effects, the program is not perceived to be relevant. Third, some wastewater farmers state that the issue is not of great concern to them because they perceive little damage to the community or the environment occurring as a result of wastewater use. Freshwater farmers who responded negatively to the government program also perceive the program as irrelevant since, from their perspective, they are using the irrigation water properly. The vast majority of the wastewater farmers (75%) who are willing to contribute to the improvement program suggest the contribution should be made in the form of an annual tax or water charge once just after harvest time, and should depend on land size. Accordingly, this payment scheme is used in our valuation study.

Willingness to Pay for Wastewater and Freshwater

In the WTP elicitation questions, the initial bids are ETB 20 and ETB 40 per hectare per year, randomly assigned to all respondents. As mentioned above, in the follow-up question, if a respondent said "yes" to an initial bid of ETB 20, the bid is increased to ETB 30, and if she said "no," it is reduced to ETB 15. If ETB 40 is used as the initial bid, the bid in the follow up question is increased to ETB 50 for a "yes" response and decreased to ETB 30 for a "no" response. The first row for the initial bid of ETB 20 summarises the "no" responses to that bid. The second row summarises the "yes" response to the first bid, etc. For example, for the initial bid of ETB 20, 20.4% respond "no" and 79.6% respond "yes." Of the 20.4% "no" responses to the initial bid of ETB 20, the follow-up bid of ETB 15 results in 24.5% "yes" responses and 75.5% "no" responses.

Econometric Results for Wastewater and Freshwater Farmers

Out of our sample total of 415 farmers, 408 responded to our valuation questions. However, due to item nonresponses for some socioeconomic variables, only 372 households were included in the final econometric analysis, 223 of whom are wastewater farmers and 149 use only freshwater for irrigation. Following the earlier discussion on econometric specifications, we estimated three different models: a standard probit model was estimated using only the responses to the initial bid (single-bounded dichotomous-choice, SBDC); the full responses to the double-bounded questions were estimated using a bivariate probit (BV-DBDC) and the interval-data model (ID-DBDC). In both the SBDC and BV-DBDC, the dependent variable is a binary response to the valuation question, in which respondents are given a value of one if they respond "yes" and zero if they respond "no" to the first bid. In the double-bounded model, we have another dependent variable where respondents are given a value of one if they respond "yes" to the second bid and zero if they say "no."

Figure : *Missouri Beginning Farming: Freshwater Prawn Production in Missouri*

For the wastewater farm area the number of years with irrigation experience, education, total annual yield value, and the kind of policy option significantly affect the probability of "yes" responses to the initial bid. Further, the location of the farm significantly influenced WTP. As expected, it was found that the longer the experience of farmers with wastewater irrigation, the higher their WTP for an improvement policy. This effect is plausible since those farmers are more familiar with wastewater being a valuable resource for their agricultural production and also know of its risks. This variable is not significant in the freshwater farm areas where irrigation is only a recent phenomenon.

Moreover, the less educated households seem to care less for such policies, as indicated by a significantly negative relationship of the education category "maximum of four years of schooling" with WTP. It can be hypothesized that these households have difficulties in perceiving the environmental and health problems arising specifically from the contaminated irrigation water, and consequently consider the improvement policy less useful. In contrast, as expected, income (total annual yield value) is positive and significant, implying that those farm households who obtained higher yields benefit more from an improved and safe wastewater irrigation system, and therefore states a higher WTP. It is quite unexpected, however, that in the ID-DBDC model the kind of policy option to make wastewater irrigation safer has a significant effect on WTP. Respondents appear to be less willing to contribute to the "regulation enforcement" policy than to the "awareness creation" policy. This may be due to the notion that it is essentially the government's task and responsibility to enforce existing regulations and prosecute violators. However, this variable is not significant in either the SBDC or the BV-DBDC models.

Use of irrigation water for purposes other than crop production, and households' perception of the benefits they receive from irrigation water have no significant effect in the ID-DBDC model, though both variables had a positive sign. Perceived benefits from irrigation water are significantly positive only in the BV-DBDC model for wastewater irrigators but not for freshwater irrigation. This finding implies that, as expected, wastewater farmers attach a positive value to the wastewater. The variable for the water extraction method has no significant effect on the probability of a positive response to the first bid. Therefore, other factors remaining the same, the farmers' WTP for the improvement program is not related to the way the water is extracted from its source. Consequently, to improve the safety of wastewater crop production, it is possible to introduce water saving and improved irrigation methods such as drip irrigation without affecting farmers' utility from irrigation. The dummy variables for location have significant effects, and their signs in the separate regressions are as expected. Farmers working on upstream and middle-stream of the wastewater farm areas have a higher WTP than those downstream. This is explained by the fact that the middle and upstream farmers harvest at least twice per year and obtain higher annual yield value than the downstream farmers, the majority of whom are producing only one harvest per year due to water-logging problems on their farms. Farmers' perceptions of the health effect of irrigation are insignificant in the model, revealing that farmers attach a positive value to the wastewater regardless of the health risks, perhaps because its use as a means of livelihood outweighs the perceived risks.

In our fitted bivariate probit model for the wastewater farm area, the estimate of the coefficient of correlation of error terms <1 of the double-bounded model is 0.88 and statistically significant at the 1% significance level. This confirms the theoretical assumption that there is a strong positive linear relationship between the random components of the responses to the initial bid and the second bid. The estimate of 0.88 is rather close to perfect correlation, and consequently indicates respondents largely drew their responses to the first and the second bid from the same underlying WTP. Further, this high value of <1 validates our selection of the interval-data model as opposed to the more general bivariate probit model, since the bias of parameter estimates is only small at such a value of <1 while the statistical efficiency of the interval-data model is much higher (cf. Alberini, 1995).

Single- and Double-Bounded Models: A Comparison

Mean and Total WTP

The ultimate aim of fitting a statistical model to CV responses is to derive a summary measure of the WTP distribution (mean WTP) and to estimate the welfare change to society due to the improvement program. The WTP results, in ETB per ha per year, were estimated assuming a linear random utility model. The result shows that mean WTP from the single-bounded model is ETB 39.57 per hectare per year, and for the bivariate probit model is 39.10, which is only marginally smaller. The mean WTP from the intervaldata model is 39.72, which is also similar to the mean WTP from both models, and thus in line with Alberini (1995). As expected, the mean WTP from the open-ended follow-up question is slightly smaller than for the other models since respondents have a tendency to state their maximum WTP in close vicinity to the highest accepted bid. The mean WTP of wastewater farmers is higher than that of the freshwater farmers based on the openended and interval models, though this difference is not significant.

From an economic policy perspective, this mean WTP indicates that an average farm household is willing to pay 0.37% of its annual farm income for the improvement program. Comparing the results from the two groups of farmers, an average wastewater farm household is willing to pay 0.27% or

0.28% of its annual average farm income per hectare, based on BV-DBDC and the interval-data mean WTP, respectively. The corresponding figures for the average freshwater irrigator household are 0.69% and 0.65% of its annual farm income per hectare. This means that taking the average farm land size in the wastewater area (0.4 ha) and average annual farm income of ETB 3,146.20, the average wastewater farm household is willing to pay 0.5% of its annual farm income per year for the improvement program. Given this low share of WTP as compared to income, we deduce that in our study no systematic underestimation of benefits resulting from restrictive budgets did occur.

The total WTP (aggregate benefit) is the total economic benefit that can be obtained from the hypothetical improvement program for the safe use of irrigation water in agriculture. Mitchell and Carson (1989) discuss a number of issues that should be considered in benefit aggregation. These are population choice, sampling frame, sample nonresponse, and sample selection biases.

Accordingly, the potential individuals who would benefit from the improved program in our study areas are the farm populations. The farmers are those who receive the direct benefit from the improved service and pay for the services. Therefore, we estimated the total benefit for the total farm population from which our sample is drawn using information on the total farm lands used for irrigated agricultural activities, obtained from the local agriculture offices. Unit nonresponse is not a problem in our survey as all farm households who were asked for an interview agreed to participate. Still, item nonresponse was observed, since out of the 415 farm households who filled out the questionnaire, only 408 completed the entire questionnaire; i.e., 1.7% of the farm households did not respond to the valuation question even if they answered the remainder of our questions. This is very low compared to what is generally expected from any CV survey, and would not result in selection bias if excluded from estimation. Of the 408 farm households who responded to all questions, 1 5 responded zero WTP. We asked a follow-up question as to why the respondent was not willing to pay any amount so as to identify whether such responses are protest zeros or true zeros. None of the responses were found to be protest zeros, as the respondents explained that they cannot afford to pay any amount.

The total WTP of the wastewater and freshwater farm population in the study area for the policy programs investigated is estimated at ETB 92,964.66 per year based on the interval-data mean WTP, and ETB 91,513.55 based on the bivariate probit model. From the open-ended mean WTP, the total welfare gain is estimated at ETB 82,736.68. The highest welfare gain is observed based on the estimate of the interval-data model. Note that we found is significant and rather high in the bivariate probit model estimation, which implies interval-data estimation is more efficient. As discussed above, the lowest welfare gain is obtained from the open-ended mean WTP. The welfare gains to the wastewater farm community from improved use of wastewater for irrigation. It ranges from ETB 44,979.42 to 49,41 1.32 per year based on the open-ended and interval-data estimations, respectively. These findings can be compared with the welfare gain from the improvement program in the freshwater area, which is lower in all estimations.

Validity and Robustness of Results

The validity and robustness of the results are checked internally in light of the standard economic theory and by making use of specific indicators mentioned earlier.' We have already explained that the design and implementation of the survey was conducted to minimise the biases or problems that can arise when using CV methods, including scenario specification, elicitation questions, and payment vehicles, as well as in procedural matters including sample design and analysis of the data. Therefore, our CV study passes the content validity assessment. We also assess the robustness of our study based on construct validity in which we test whether the important explanatory variables are in line with what economic theory suggests. In the context of our study, we can hypothesize that the farm household's WTP increases with higher income from farming. As shown in the regression result, income, which indicates the household's ability to pay, increases the probability that the respondent will accept the offered bids, as well as the willingness to pay. The share of the amount of money farm households are willing to pay from their total household income, about 0.37% on average, is also quite reasonable. Moreover, consistent with the theory of demand, the coefficients of the first and second bid are negative and significant at the 1% probability level, indicating that a higher bid decreases the probability of accepting the offered bid.

As detailed previously, the application of the CVM in the socioeconomic setting of developing countries poses particular challenges. These may arise especially from limited cognitive capabilities, lack of trust in government, and low monetary incomes. We address these issues and conclude that the CVM is in principle applicable in the context of our survey population.

First, all household heads in our sample live in the vicinity of Addis- Ababa and received a minimum of formal education. However, the understandability of the relevance of the problem with wastewater irrigation and the policy options to reduce these were tested within various rounds of focus groups. This process allowed for the design and formulation of the questionnaire and valuation scenario as well as the payment scheme that is adapted to the cognitive capabilities of the survey population. Here, it was helpful that farmers have had a longtime experience with wastewater irrigation and are very familiar with its benefits and problems.

Second, it was found in our survey that, while the majority of respondents chose a combination of policy options 1 and 2 (awareness creation in combination with enforcement of existing regulations), WTP was significantly lower for the latter. From this finding we conclude that respondents are sensitive to the particular policy option presented in the scenario, and thus reflect on their willingness to contribute to such policies in a rather differentiated way.

Third, the shares of stated WTP to the households' incomes were found to be sufficiently small (as is typical in developed countries) to conclude that their WTP is not restricted by their ability to pay, and therefore not downward biased. This interpretation is underscored by the finding that WTP is not negatively influenced by the number of dependents, which would be a typical indicator for tight family budgets. From these indicators, we conclude that the prerequisites for employing the CVM as an economic method to value the social benefit of environmental or nonmarket goods in the socioeconomic context of urban and peri-urban farmers in Ethiopia are satisfied.

This chapter has examined the resource value of the safe (improved) use of wastewater in crop production within and around Addis Ababa, Ethiopia, based on farmers' willingness to pay. The potential welfare gain from a government policy to provide safe wastewater use was estimated employing the contingent valuation method (CVM). We also compared the value of the improvement program of wastewater with a comparable program for the improvement of freshwater use in irrigation. The double-bounded dichotomous-choice elicitation question format was applied to elicit farmers' value of the improvement program. In-person interviews were conducted at 415 farm households operating irrigation in wastewater and freshwater areas. Responses were analysed using descriptive and econometric techniques.

The descriptive results indicate that wastewater is used for multiple purposes including domestic uses and irrigation. For the majority of wastewater farm households, the wastewater is the only means of survival, and the average income from wastewater farming accounts for 62%) of total annual household income, with this value ranging from a low of 27%) to a high of 97%). Accordingly, 88% of the wastewater farm households report that they benefit from the wastewater. Yet, farmers also perceive that the wastewater poses health risks for their families and for consumers. Likewise, they see negative externalities to the environment as it pollutes the soil and causes foul odours in the surrounding community. These negative externalities are more pronounced in the downstream wastewater farm areas where the pollution levels are higher and the benefits from wastewater are relatively lower.

Among the options provided to the wastewater farmers for improving the existing situation, "enforcing existing environmental regulations and prosecuting violators" is the most preferred option, with nearly 40%> of the farmers selecting this choice. "Awareness promotion on safe use of wastewater" is another option selected by the farmers to improve the situation. This includes technical advice and training on improved irrigation methods, crop selection, safe consumption of vegetables, and quality protection, as well as access to protective clothing at affordable prices. Farm households are willing to contribute to the improvement programs proposed to them, and the payment vehicle could be made in the form of a mandatory annual tax just after harvest time, depending on the land size.

The WTP of farm households and its determinants were analysed under different econometric specifications: the single-bounded probit, the double-bounded bivariate probit, and interval-data models. Comparing these specifications, we obtain robust results. The results show that, as expected, the number of years with irrigation experience, education, total annual yield value, and the kind of policy option significantly affect wastewater farmers' WTP. Further, the location of the farm significantly influenced WTP. Since we found that the coefficient of correlation of the error terms of the double-bounded bivariate probit model is close to 1 (almost perfect correlation) and statistically significant at the 1% significance level, we suggest using the interval-data model to achieve more efficient estimates of the WTP for the improvement program.

From our models, we conclude that wastewater farmers tend to be willing to pay more than freshwater farmers for an improvement policy. Additionally, as expected, wastewater farmers who stated they benefited from the wastewater are willing to pay more than those who did not perceive much benefit. The results also indicate that the middle and upstream farmers are willing to pay more than farmers in the downstream areas. This could be due to the water-logging problem in the downstream areas that limits the downstream users to only one harvest per year. The variation in pollution levels is also reflected in farmers' responses to the foul environmental odour caused by the wastewater, on which the downstream farmers strongly agree but those from the upstream areas do not. This result implies that the value of the improvement program is also explained by variations in the pollution levels between the different wastewater farm areas.

The mean WTP of wastewater farm households from the bivariate probit and the intervaldata models is estimated to be ETB 37.15 and 39.88 per hectare per year, respectively. Based on this mean WTP, the welfare gain for the wastewater farm populations obtained via the safe use of wastewater for irrigation ranges from ETB 46,028.85 to 49,41 1.32 per year, respectively.

Overall, the results presented here confirm that safe use of wastewater for irrigation, or improved use of the irrigation system, is important to farm households and that the welfare gain to the farm population is tremendous.

It also confirms that wastewater is a valuable resource for the livelihoods of farmers. Both the wastewater and freshwater farmers are willing to pay for programs designed to improve the existing unsafe and improper use of irrigation water. This finding suggests it is possible to introduce an irrigation water users' fee that could signal scarcity in the quantity and quality of irrigation water and could optimise or at least improve efficiency in the use of wastewater or freshwater irrigation water.

The study not only provides useful estimates of farm households' WTP for safe use of wastewater, it also offers guidance for decision makers to make informed decisions concerning how to maximise the benefit from the use of wastewater for crop production while minimising health and environmental risks. It provides an important option for the existing treatment plant to improve its services by selling treated (or semi-treated) wastewater to farmers for a fee, since farmers are willing to pay for such an improvement program.

From a methodological perspective, we conclude that the CVM is suitable for use in the socioeconomic setting of rural-urban farmers in a developing country like Ethiopia. In this survey very little protesting behaviour was found, which we attribute partly to providing the respondents with a variety of policy option choices, all leading to the same safe use of wastewater for farming. Further, the robustness of the WTP estimates across various econometric specifications for the dichotomous-choice format shows that typical problems with this format, such as anchoring effects, were not influential in our survey.

However, our investigation is limited to the perceptions of farm households, and the results are subject to this limitation. We recommend further studies on the attitude and willingness to pay of the residents of Addis Ababa for improved waste management services, as they are both sources of wastes and consumers of wastewater-irrigated vegetables. We also suggest further analyses should be conducted to determine the optimum level of treatment on industrial wastewater and to develop incentive-based mechanisms to treat the wastewater before discharging into the rivers.

7

Wastewater Treatment in the 21st Century: Technology and Regulatory Issues

As we prepare to enter the 21st century, it is time to evaluate our methods of on-site wastewater management. The evaluation should address not only the tools of the trade (i.e., on-site wastewater treatment and disposal technologies), but also the process in which the technologies are used (i.e., the operation and management of the systems and the regulatory framework). Advances in technologies for small-scale collection, treatment, and disposal, as well as the availability of remote monitoring systems, now make it possible to offer the most advanced wastewater treatment options in low-density areas at a cost lower than that of conventional large-scale pipe-and-plant systems.

The reasons are as follows. Generally, in small communities, houses are spread out, and the density is low, which makes the use of an on-site system for an individual home or for a cluster of homes a cost-effective option. Wastewater management systems can be engineered to minimise the collection cost as far as possible, typically to le ss than 30 percent of the total cost, with currently available on-site wastewater treatment and dispersal technologies. Because there are now available a variety of pre-engineered, prepackaged treatment systems for small flows, it is not necessary to collect a large quantity of raw wastewater at one point for treatment.

Thus, minimising the collection network for raw or partially treated wastewater can minimise the total cost of any wastewater project in any small community, new or existing. The total cost includes the operation and maintenance costs for the expected life of the wastewater system.

For most of the 20th century, a standard septic-tank drainfield system has been the primary means of on-site wastewater management. Census data from 1990 estimate that more than 25 million homes in the country are not connected to a central wastewater system and use some form of on-site system, most likely a septic-tank drainfield system. Probably the first advance in this technology was the use of a pump to overcome gravity when a "suitable" drainfield site was at a higher elevation than the house. A conventional septic system uses soil to treat primary or raw wastewater discharged from a septic tank.

Research studies have documented how a drainfield works at a given point in time on a given site; the real impact millions of drainfields have on the environment and public health, however, is not adequately documented. Conservative rules for locating and sizing septic systems as specified in state or local health department regulations are the primary reason no major environmental or public health problems have been associated with the use of millions of septic systems so far. No one really knows what kind of treatment a subsurface drainfield actually achieves on a long-term basis. It is hard to collect effluent samples below a subsurface drainfield. Thus, it is not possible to monitor the performance of such a system. The performance of a septic drainfield system is taken for granted as long as conservative siting and design standards are followed during installation.

As it becomes clear that on-site systems will be used on a permanent basis and will be needed in areas not suitable for treating primary or raw sewage (i.e., land that does not percolate), we must look for systems that treat wastewater to secondary or higher quality before discharge. We also need site-assimilative systems for safe dispersal, recycling, or reuse of treated effluent. Finally, we need an infrastructure to operate and maintain small systems on a permanent basis, and we need a regulatory framework that allows people to manage on-site systems in a cost-effective and environmentally sound manner.

Technologies

Wastewater can be managed in small quantities with a variety of technologies. Technologies are available for collection and treatment of raw wastewater, as well as for disposal, dispersal, recycling, or reuse of treated effluent. Since the small systems are typically spread out over a large area, day-today monitoring becomes a real challenge. Fortunately, we now have access to telemetry systems (i.e., remote monitoring), which can operate small systems and send the performance information to a central computer by telephone line or by other means of communication. Thus, the tools are available to manage wastewater on site in small quantities in an environmentally sound and cost-effective manner. Widescale use of these tools, however, is not possible, mainly because maintenance infrastructures are lacking and the regulatory framework is inadequate. Total cost of wastewater systems, both the capital cost and the costs of permanent operation and maintenance, can be minimised with on-site technologies. The capital cost typically includes cost for collection, treatment, and disposal, along with the cost of engineering and construction management.

Collection System

The main reason for collecting raw or partially treated wastewater from more than one structure is to reduce the cost of the treatment and disposal system. At the same time, a cost is associated with the installation and operation of the collection system itself. Thus, one needs to determine the optimum amount of collection system that results in the minimum total cost of the wastewater management system. Typically, in small communities, the housing density is low and variable. The downtown business district may have commercial structures very close to each other, while the residential neighbourhoods may have houses on large lots. The total cost of the wastewater system can be minimised if a number of small treatment and disposal systems are used instead of a collection method that uses one treatment and disposal system for the entire community During the planning phase, one needs to analyse the costs of various options, starting with the bare minimum amount of collection to the use of a collection system for the entire community.

Wastewater may be collected in raw form, typically with a gravity or vacuum collection system, or it may be collected as effluent from a septic tank or grinder pump, typically with small-diametre pipe pressure or a gravity system. The Alternative Wastewater Collection Systems Manual, published by the U.S. Environmental Protection Agency (U.S. EPA) gives details on small-diametre gravity, pressure, and vacuum collection systems. It also compares these alternatives with the conventional collection approach of gravity sewer and pump stations. It is a challenge to wastewater engineers first to select an appropriate technology for collection and second to optimise the level of collection system such that the total cost of a project is minimised. Sometimes excessive collection infrastructure is viewed as an investment toward a future need for wastewater services. Such an approach could, however, easily lead to unnecessarily higher capital cost that may never be recovered if the expected demand does not materialize . Instead of spending resources for a collection system that may not be needed in the foreseeable future, small communities should address their current needs first and establish a management infrastructure—a utility—that can adequately address additional needs if they arise.

Alternative collection systems such as septic-tank effluent gravity systems (STEGs), septic-tank effluent pump systems (STEPs), grinder pumps (GPs), and vacuum systems typically use components that require routine operation and maintenance, just like pump stations in a conventional system. For example, the solids collected in septic tanks (interceptor tanks) may need to be removed, the pumps (effluent or grinder) may need to be replaced, screen pump vault or effluent filters may need to be cleaned, or the pneumatic valves used in vacuum sewers may need maintenance or replacement. Adequate access to such components must be possible at all times, and some type of monitoring tool must be used to watch the performance of pumps or other electromagnetic components and liquid levels in the tanks. All such services must be performed by a management entity that is responsible for operating the entire wastewater system. The cost of installing, operating, and maintaining the collection system should be weighed against the savings in treatment and disposal systems.

Treatment Systems

The most important objective of any wastewater system should be to treat the wastewater adequately before it is released, recycled, or reused. Thus, the treatment component of a wastewater system must be appropriately addressed and funded. The scientific principles of wastewater treatment are well researched and documented, and they are taught in engineering schools, A number of companies manufacture a variety of on-site systems that treat wastewater in small quantities, typically less than 1,000 gallons per day. New concepts and systems are being developed. Nevertheless, the vast majority of structures that are not connected to a central sewerage system today use a very simple and passive wastewater treatment system called a septic-tank drainfield system.

The septic-tank drainfield system has two components—a septic tank that separates heavy solids and light material from raw wastewater and a subsurface drainfield, which is a series of subsurface trenches. This system depends primarily on soil surrounding the trenches for treatment of the septictank effluent. Since septictank effluent—primary effluent—has high levels of organic matter, a bio-mat is formed at the soil and gravel interface in the trenches. The main idea behind the passive system is that a certain quantity and quality of unsaturated welldrained soil—along with the bio-mat-can treat septictank effluent to the degree necessary to protect groundwater or surface water present in the area.

The soil acts as a medium and offers biophysical treatment of septictank effluent. Since the drainfield is installed underground, there is no mechanism for inspection or maintenance of this treatment system, nor is there an effective way of monitoring the effluent quality Performance of septic systems is taken for granted as long as certain predefined soil and site conditions are present on the site, the system is sized according to loading rates specified in the regulations, and the specified distances separate the system from sensitive areas. State or local health department regulations prescribe soil and site conditions, along with design standards necessary for use of septic systems, in terms of depth and type of suitable soil, permeability or percolation rate, and distance to surface water, groundwater, and other environmentally sensitive areas, Using these prescriptive regulations, a sanitarian or a licensed soil evaluator typically performs soil and site evaluations, and the proposed site is either accepted or rejected. Rejection of a site usually means the lot is condemned as far as habitable-building purposes go, even when it is good in all other ways for development. At the same time, acceptance of a site based just on soil and site characteristics should not be considered assurance of long-term environmental protection.

The idea of treating wastewater to a secondary or better quality before discharging effluent into soil is drawing a lot of attention. This may be due to the fact that we are running out of sites that are considered "good" for septic systems, or it may be due to the realisation that we can do better. Today, a number of pre-engineered, prepackaged small wastewater treatment systems allow subsurface dispersal of effluent on any site that limits the use of septictank effluent. These treatment systems also should be used for sites that are considered "good" for septic systems—just as it makes no sense to discharge primary effluent into surface water, so it also makes no sense to discharge primary effluent into the subsurface environment. With the availability of small on-site treatment systems, any site that has potential for a residential or a commercial building can have a wastewater system, provided the owner is willing to pay the costs associated with the engineering, installation, and operation of such a system. When a connection to a central wastewater system is not available, a wastewater solution can be found for any area through the use of an adequate nonseptic on-site system. Raw wastewater or septictank effluent can be treated to any of the following levels: secondary, advanced secondary (secondary treatment followed by disinfection), or tertiary (secondary treatment with significant nutrient reduction and disinfection). The treatment technologies can be grouped into five categories:

1. aerobic treatment units (ATUs) comprise the following technologies:
    a. suspended growth—flow-through or sequencing batch reactor,
    b. attached growth—trickling filter with forced aeration, and
2. media filters—single-pass or recirculating—can use a variety of filters, including
    a. a combination of suspended and attached growth;
    b. sand filters,
    c. peat filters,
    d. foam filters, and
    e. textile filters;

3. natural systems for polishing or recycling of secondary effluent include
   i. wetlands and
   ii. greenhouse systems;
4. waterless toilets and graywater systems (as alternatives to flush toilets) may take the form of
   i. composting toilets or
   ii. incinerating toilets; and
5. disinfection systems for secondary or higher-quality effluent include
   i. ultraviolet light and
   ii. chlorination and dechlorination.

Aerobic Treatment Units

ATUs offer an alternative to septic tanks, and they treat raw wastewater to secondary treatment standards (BOD,//TSS [less than] 30 mg/L). In some ATUs, a trash tank, for primary treatment, precedes the aeration tank. A number of pre-engineered ATUs are available, and they are generally used on sites that soil and site evaluations find unsuitable for a septic drainfield system. To evaluate the performance of small ATUs, NSF and Baylor University have a testing-and-certification facility that applies the ANSI/NSF Standard 40 for Class I effluent limits.

Typically, the effluent from ATUs, after further polishing and disinfection, is discharged into a surface-water body or on top of the ground, resulting in a point-source discharge into an adequately sized subsurface system instead of a nonpoint-source discharge.

Subsurface disposal of secondary-quality effluent is technically possible on sites that are not suitable for primary effluent (i.e., septictank effluent). Actually, subsurface dispersal of secondary effluent by methods such as filter beds or drip/spray systems can have less impact on the receiving environment than surface discharge.

Plants can take up nutrients such as nitrogen and phosphorus when the secondary effluent is dispersed into ground at a shallow depth, typically within the top 12 inches of soil. Sites at which the soil has low permeability, at which the depth to an impermeable layer is shallow, or at which the depth to seasonal groundwater is shallow can be used for subsurface dispersal of effluent from adequately operating ATUs.

The ATUs must be operated and maintained by professional operators to produce high-quality effluent on a permanent basis. When not adequately operated and maintained, any ATU—or, for that matter, a septic tank or any other treatment system—will discharge inadequately treated effluent into the environment and cause problems.

Media Filters

Used primarily to treat septictank effluent, media filters sometimes also are used to polish effluent from ATUs. The most commonly used media filters are sand filters (single-pass or recirculating). Recently, other types of media, such as peat, synthetic foam, and textiles, have been evaluated and have been shown to treat septictank effluent to better-than-secondary quality As with ATUs, we now have access to pre-engineered, prepackaged media filters that can be easily installed and used for advanced treatment of septictank effluent. Instead of engineering a media filter or an ATU for individual small applications, it is advisable to obtain such treatment devices from companies that market them and to use engineering resources to develop subsurface dispersal systems that will minimise environmental impact. After adequate disinfection, effluent from both the media filters and the ATUs also can be recycled for flushing toilets or other nonpotable uses, thus reducing the need for subsurface dispersal.

The performance of any media filter will depend on the quality of the media, the recirculation rate, the volume of the recirculation tank, the kind of distribution system used to spread effluent on top of the media, and the ventilation of the media filter itself. Use of an effluent filter in a septic tank and regular maintenance of the septic tank also are necessary if the media filter is to perform adequately. It may be necessary to change media filters of some types after a certain number of years. Media filters have been demonstrated to be very effective at reducing organic and bacteriological contaminants from septictank effluent. They also can convert most of the nitrogen to a nitrate form (nitrification), thus maximising the potential for plant uptake if the effluent is adequately dispersed into the shallow root zone with shallow-trench dispersal, drip, or spray. There is a potential for a high degree of denitrification when recirculation through the septic tank is adequately designed.

Natural Systems

Wetlands and other natural systems are being used to treat septictank effluent from single-family homes or—mainly in southern states—from a community. The performance of wetland systems depends on the design, the vegetation used in the wetland, the climate, and the operation of the system. Such systems tend to require large land areas and typically have little or no mechanism for adjusting performance according to variations in the inflow quality. At the same time, this kind of treatment has much lower energy requirements than do other options. An adequately engineered wetland system can offer a cost-effective method for removing nutrients and other constituents from primary or secondary effluent. The use of a greenhouse system, a wetland operated in an enclosed and controlled environment, can lower or eliminate dependence on climatic conditions. Greenhouse systems offer a reliable treatment mechanism that can produce high-quality effluent on a consistent basis. They also can be used to significantly lower or eliminate discharge of effluent into the environment by using plant uptake and evapo-transpiration as the primary mechanisms for assimilating effluent. Most of the time, natural systems provide cost-effective further treatment of secondary-quality effluent, discharge from ATUs, or discharge from media filters. They serve to further reduce the impact of nutrient and bacteriological pollutants on the environment.

Waterless Toilets

Composting or incinerating toilets can be used to reduce the quantity of wastewater generated from a facility and also to influence the quality of wastewater. One must, however, deal with the residual products from such a facility—either composted material or ash by recycling compost in the yard as a fertilizer or by sending ash to a landfill. The remaining wastewater is called "graywater," and it can be treated adequately with natural systems such as wetlands—or with other systems—prior to subsurface dispersal. Typically pollutant loads in graywater are greater than those in the effluent from a well-maintained media filter or an ATU; hence, adequate treatment and disposal of graywater must not be overlooked. Use of waterless toilets is attractive for remote, nonresidential areas (e.g., golf courses and rest areas in parks), where access to water and wastewater facilities is costly.

Effluent Dispersal Systems

Small wastewater treatment systems, just like large ones, need a mechanism for discharging treated effluent. As noted earlier, subsurface dispersal (nonpoint-source discharge) is the primary mechanism for disposing effluent from small on-site treatment systems. The three major parameters that influence the performance of subsurface dispersal systems are effluent characteristics, method of application, and soil and site characteristics. Of these three parameters, the first two are more manageable than the third. Highly treated effluent, when applied in small and frequent doses (this approach is called time dosing), can be adequately dispersed in a variety of soil and site conditions. The following technologies are available for subsurface dispersal of effluent from small treatment systems:

- trenches—gravity- or pressure-dosed, with or without gravel;
- drip—at or below grade;
- spray—above ground;
- filter bed—raised system on a sand-lined bed;
- evapo-transpiration beds; and
- greenhouse with storage tank.

Since most of the disposal technologies listed above use soil as a media for receiving the partially treated effluent, soil evaluation has been an integral aspect of on-site wastewater systems. A variety of treatment systems, however, now make it possible to achieve the necessary degree of treatment outside the soil. Thus, soil and site evaluation—as done for septic systems—may not be necessary In particular, there needs to be a change in the regulatory requirements for conducting soil evaluation prior to installation of a subsurface dispersal system when nonseptic on-site treatment systems are used. A variety of systems for dispersal of less than 1,000 gallons per day of treated effluent can be pre-engineered, installed, and operated in a manner that will ensure adequate assimilation of treated effluent within the zone of influence specified for that system. The site characteristics and environmental sensitivity of the proposed location will determine the appropriate type of pre-engineered dispersal system. Effluent dispersal systems should be selected and sized according to the assimilative capacity of the site relative to design flow and nutrient loading—not according to soil characteristics. Soil and site conditions that are viewed as limitations for septic drainfields must not be viewed as limitations for the use of effluent dispersal systems.

The ability of soil and site to effectively absorb and move the effluent away from the disposal site with a minimum movement of pollutants is the main issue of concern when highly treated effluent is discharged into a subsurface disposal system. Effluent dispersal systems must be viewed as site-assimilative systems, not just as soil absorption systems. For given soil and site conditions, an appropriate effluent dispersal system can be selected, installed, and operated to prevent the following occurrences:

- a point-source discharge (i.e., a stream flowing out of the area where the system is installed);
- a public nuisance (e.g., a puddle of water on or around the area where the system is operating);
- an obvious or even a perceived health hazard from the system; and
- groundwater or surface water contamination from organic, inorganic, bacteriological contaminants discharged into the system.

The above-mentioned treatment and dispersal technologies can be grouped according to wastewater system types. Within each type a number of engineering designs and methods can be used to achieve the necessary performance goals on the sites where the systems are used:

- System Type I—conventional gravity septictank effluent drainfields,
- System Type II—pressure-dosed septictank effluent drainfields,
- System Type III—drip dispersal of septictank effluent,
- System Type IV—waterless toilets and graywater systems,

- System Type V—shallow gravity trenches for secondary or better effluent,
- System Type VI—shallow pressure-dosed trenches for secondary or better effluent,
- System Type VII—drip dispersal for secondary or better effluent,
- System Type VIII—filter beds for secondary or better effluent,
- System Type IX—evapo-transpiration beds for secondary or better effluent,
- System Type X—spray dispersal for advanced-secondary effluent, and
- System Type XI—greenhouses for advanced-secondary effluent.

Two parameters that have a major influence on the functioning of a subsurface dispersal system are the depth of soil to limiting conditions (impermeable layer or seasonal water table) and the permeability of soil (the percolation rate, the hydraulic conductivity, or the texture/structure).

Sites can be grouped into four types according to the values of these two parameters. Within each site group, parameters such as slope, vegetation, and environmental sensitivity will determine the types and designs of appropriate wastewater systems. Most state regulations specify the soil depth and permeability required for a septic drainfield. With these limits as reference points, a four-quadrant matrix can be developed.

Remote Monitoring System

Most of the wastewater systems mentioned above use electromechanical devices such as pumps, blowers, and remote-monitoring float switches to achieve treatment and dispersal goals. The performance of treatment and dispersal systems depends heavily on the reliable operation of these devices.

A remote-monitoring telemetry system should be used to adequately monitor the operations of these devices on a continuous basis and to detect any problems quickly. Control panels now available for small systems are capable of operating the electromechanical devices as well as of reporting the conditions of these devices to a central computer over telephone lines or by other means. Such systems can send signals to a central computer on a routine basis or to an operator during an emergency situation, reporting information on various parameters such as pump run time, duration of power outage, high-water conditions, and so forth. By using such remote-monitoring systems, a wastewater utility can operate a number of small waste water systems installed over a large geographical area in a cost-efficient manner.

Operational information gathered from remote-monitoring systems can be used to prepare routine reports on the performance of the wastewater systems and to determine the amount of wastewater managed. Such reports can be useful for billing the user as well as for supplying information to the regulatory agency responsible for ensuring adequate operation. The information also can guide planning for efficient, routine replacement of pumps and other devices, thus minimising or preventing serious out-of-compliance situations.

Operation and Management

Centralised wastewater treatment plants are operated by a utility, public or private, whose trained and licensed operators monitor and maintain the plant so that discharge meets performance standards. Basically homeowners and businesses pay a hook-up fee to connect to a centralised system, then pay a regular usage fee, transferring all responsibility for their wastewater to the utility. Today most people who use small on-site systems do not have this option. Public acceptance of small on-site systems can be enhanced only when such systems offer the same wastewater services as a centralised system. When an on-site system can offer operational comfort to users and an environmental protection guarantee to the regulators, it will be considered the equivalent of a centralised system. We now have technologies that can meet both of these requirements in cost-effective ways. Making those technologies available to people, however, requires an infrastructure similar to that of a utility and that infrastructure is still in its infancy.

Today a few management entities offer wastewater services to people who use small on-site systems; however, most people still have no access to such services. At the beginning of a new century we need to think seriously about how to develop a regulatory system that will allow people to obtain wastewater services from a utility that uses small on-site systems in the same way they get other services—solid waste, telephone, cable, gas, and power.

We also need to discuss what kinds of services utilities should offer and what role such companies could play in the on-site industry. When a utility is responsible for permanent operation and maintenance of an on-site system, it is possible to address simple issues, such as access to system components for maintenance and inspection, in a timely manner. The current regulatory requirements of soil and site evaluation, engineering design, and multiple reviews should not apply to a licensed utility. A qualified and licensed utility should be permitted to do all the pre-installation work (e.g., engineering, site and soil evaluation, selection of a wastewater system) and should be allowed to install and operate on-site systems. The utility also should be allowed to use the best technology available for wastewater treatment and dispersal, and should be regulated according to the performance of the systems, in terms of both operational services to the customers and protection of the environment and public health. Under the utility model, the roles of manufacturers, engineers, soil and site evaluators, and installers can be defined in a manner that would promote the most efficient use of their services. Today the requirements of soil and site evaluation and of engineering often do not add any real value to the operation of individual home and small commercial wastewater systems. Most of the current regulations for on-site systems still require soil and site evaluation to determine if the proposed site is suitable for an on-site system. Such pass/fail criteria for a site are not necessary because it is now possible to have a wastewater system for any site. Once a decision is made for development in an area that is not served by a centralised wastewater system, an on-site system utility can offer all the services necessary for adequate treatment and dispersal of wastewater. The environmental and public health regulators then can ensure that the services offered by the utility provide safe, adequate, and proper protection of the environment and public health by making sure that the utility is using the best technologies available for wastewater treatment and dispersal and by monitoring the performance of the on-site systems and their impact on the environment.

A utility company can help the on-site industry weed out poorly designed or poorly manufactured wastewater technologies. At present, there is no real mechanism for measuring the long-term performance of small wastewater treatment and dispersal systems. A utility company responsible for acquiring, installing, and operating a wastewater system in a manner that meets performance standards and satisfies customers in a cost-effective way will always strive to use the best possible technology.

A utility company also can educate people about the environmental impacts of wastewater and about the importance of reusing or recycling adequately treated wastewater. There is tremendous interest in the use of environmentally friendly systems and the reuse of treated wastewater. One must, however, realise that improperly managed wastewater systems can create environmental and public health problems. Only under a proper management framework should people have access to advanced, environmentally friendly wastewater systems. Finally a utility company can help people get the best possible wastewater system at the least possible cost by acquiring products and services in quantity. Today most people who apply for an on-site system permit—typically to a health department—get most of the pre-installation services (e.g., soil evaluation and design) from a health department employee, usually a sanitarian.

These employees often are trained in only one type of on-site system—the septictank drainfield system. When it is determined, however, that soil and site conditions are not suitable for a septic system, the homeowner is asked to retain the services of the private sector for help with alternative systems and has to purchase the products and services necessary for installation of those systems. Thus, the current regulatory system is the main reason there are so many septictank drainfields in the country and so few alternative on-site systems that treat wastewater to secondary or higher quality before discharge. The process that could establish a wastewater utility model in a state must start with changes in legislation. Most important, we need legislation that sets a time frame for phasing in the use of appropriate on-site systems under the utility model and phasing out the use of conventional septictank systems.

Regulatory Framework

Government agencies that are responsible for regulating wastewater systems must focus on two important issues: (1) adequate treatment of wastewater and dispersal or reuse of effluent, and (2) environmental and public health protection from wastewater. The regulators must keep these issues in focus and develop regulatory strategies around them. The science and technologies for treating wastewater and for ensuring drinking-water quality are well established. The regulatory programs must be developed to allow wastewater professionals to operate in the competitive marketplace by offering on-site wastewater systems in a cost-effective and environmentally sound manner.

Unfortunately such a regulatory framework does not exist today for small on-site systems. The regulatory agencies responsible for on-site systems are more involved with pre-installation issues that typically have fewer direct implications for the long-term performance of a small wastewater system. For a utility to function and offer wastewater solutions to the public, the regulatory framework will have to shift 180 degrees. We need a solution-driven, performance-based regulatory framework with a heavy emphasis on post-installation issues such as system operations, monitoring of the environment, and education and training.

A solution-driven regulatory system means that if regulations are used to prescribe wastewater systems, those regulations must lead to a set of solutions for any given site and situation. The solutions should apply the best technologies available for treatment and dispersal. One way to achieve such a goal is to develop a manual of practice (MOP) for all available small-scale wastewater treatment and dispersal technologies and to update the MOP so that it stays current as technologies are developed by the onsite industry Development of the MOP must be a joint effort of the public sector; state-level technical staff; and wastewater professionals, engineers, and manufacturers from the private sector. It should include information on sizing, layout, start-up process, operation and maintenance requirements, operational cost, expected performance, zone of influence, and other issues connected to the use of the technology The MOP then can be used by a utility licensed to offer wastewater services through small-scale wastewater technologies. Technology performance data collected by the utilities can be used to revise or delete MOP content.

The utility will be the party most interested in looking at the long-term ability of a wastewater system to meet performance standards and satisfy customers at a reasonable cost.

Thus, the best source for information on the long-term use of a technology would be a utility Since at present there are no such utilities, the first version of the MOP might be based on wastewater engineering textbooks, third-party test reports, sensible ideas and claims made by engineers and manufacturers, and information gathered from U.S. EPA and from demonstration projects. Today, a utility could choose from more than 100 pre-engineered options for managing wastewater on site.

A performance-based regulatory framework should be developed from a clear understanding of how an on-site system needs to function. A widespread myth among regulators and soil evaluators is that an on-site system can work only if there is deep, dry welldrained, permeable soil (good soil) on a lot. This belief is based on a limited understanding of subsurface water movement, commonly determined by percolation or saturated-hydraulic conductivity tests or estimated on the basis of soil texture.

In reality subsurface water movement is a complex phenomenon and is very hard to predict just from observation of soil characteristics. Regulations should clearly define what types of conditions must exist on and around the area where site-assimilative systems are to be installed and operated instead of arguing about the hydraulic conductivity or percolation rate of the soil. The performance-based regulations should assign limits for effluent prior to discharge, according to environmental sensitivity and size of the system. The regulations also should assign mass-loading limits for inorganic material, total nitrogen, total phosphorus, and pollutants at the boundary of the site-assimilative system.

The boundary around a nonpoint-source discharge system can be viewed as the zone of influence for the site-assimilative system. By defining the zone of influence we can move away from regulating soil criteria, site criteria, and setback distances and allow the industry to develop new technologies that have smaller and smaller zones of influence. A recycle/reuse system, such as a system in which toilets use effluent to flush and effluent is recycled for plant growth in a greenhouse, would have the smallest zone of influence—0 feet around the greenhouse.

A lined evapo-transpiration bed may have a zone of influence of 0 feet below the system and 10 feet around the system. For any dispersal system, water quality outside the zone of influence must be no different from rain or surface-water quality allowed for public contact. Adequate penalties must be enforced when a utility violates predefined standards for effluent or mass loading of pollutants.

A performance standard also should include customer satisfaction with the overall wastewater services offered by the utility Customer satisfaction can be measured according to parameters such as sewage backup in the house, odor or noise nuisance, surfacing of effluent in the yard, and unattended alarm calls, all of which could result from inadequate operation of the systems. The performance-based regulations must indicate a method for establishing a violation and penalties for violations of each standard.

The penalties should take the form of monetary fines and revocation of licenses. Under a free-market model, an adequate number of utilities would be available to offer dependable services to all the citizens, as long as the citizens pay the fees (sewer bills) and the regulators strictly enforce performance standards. If a utility is allowed to operate while violating the performance standards, there will be no incentives to offer wastewater services that use adequate treatment and dispersal technologies.

Utilities should be informed about the expected performance standards, the methods by which performance will be measured, and the consequences of not meeting the standards. At the same time, a utility needs a legal framework that gives it authority to collect service fees and to take action against those who do not pay the fees. Small wastewater systems that allow effective recycling and reuse of adequately treated effluent on site with a minimum degree of collection will be the method of choice for wastewater treatment in the 21st century Many technologies for treatment and effluent dispersal are currently available, and new ones are being developed.

Once the infrastructure for operation and management of these technologies is established and a performance-based regulatory framework is adopted nationwide, extensive application of small-scale wastewater treatment technologies will be possible.

Nitrogen Removal from Onsite-Generated Wastewater

Conventional wastewater treatment results in the oxidation of carbon (five-day biochemical oxygen demand, or BO[D5]) and the removal of total suspended solids (TSS). Recently, the nitrification of organic nitrogen and ammonia (total Kjeldahl nitrogen [TKN]) to nitrate and the denitrification of nitrate to nitrogen gas have received much attention from the regulatory community. This attention is especially marked for onsite-generated wastewater because of an increase in population densities in areas without sewers and in environmentally sensitive regions. The U.S. Environmental Protection Agency recognises the permanence and importance of onsite wastewater treatment. Therefore, new processes that can more comprehensively treat wastewater, including the removal of nutrients and optimisation of existing technologies, are being developed.

In a conventional wastewater treatment system, BO[D5] provides a soluble substrate for microorganism growth resulting in the production of water and carbon dioxide. Nitrogen is required for growth and originates from the wastewater. With the removal of excess cells from the secondary clarifier, nitrogen contained within the removed cells exits the process. Because the nitrogen that had been incorporated in the removed cells is far less than that present in the wastewater, strict standards for effluent nitrogen levels cannot be reached simply by removal of BO[D5] and the associated excess cells from the system. Furthermore, excess cells are seldom discarded from onsite wastewater treatment as these processes operate as an extended aeration basin. In these systems, newly formed cells are allowed to oxidise to carbon dioxide and water, resulting in the release of nitrogen back into the water.

To achieve BO[D5] reduction and substantial removal of nitrogen (comprehensive wastewater treatment) to below the levels required for cell growth, multiple microbiological populations and environments are required:

- BO[D5] oxidation to carbon dioxide and water requires a heterotrophic population and aerobic conditions.
- Nitrification of organic nitrogen and ammonia to nitrate requires an autotrophic population and aerobic conditions. Before substantial nitrification occurs, the BO[D5] must be substantially depleted.
- Denitrification of nitrate to nitrogen gas requires a heterotrophic population and anoxic conditions. Denitrifiers require an easily degradable carbon source.

To achieve multiple populations and oxidation/reduction environments multiple tanks are often used. Several examples are described by Metcalf and Eddy. When sequential tanks are used for BO[D5] oxidation/nitrification and denitrification, carbon for denitrification must be added because the BO[D5] (a very good source of easily degradable carbon) has largely been depleted once the wastewater reaches the denitrification tank. There are several options for providing carbon for denitrification. Nitrate-rich effluent can be recirculated to the front of the plant, mixed with the influent water that is high in BO[D5], and then treated in an anoxic reactor (the first treatment unit). Alternatively, influent wastewater (high in BOD) can be metred to and mixed with nitrate-rich effluent and then proceed into an anoxic reactor (the last treatment unit). Instead of the influent BO[D5] being used for carbon, an external source such as methanol can be mixed with the secondary effluent and treated in an anoxic reactor (the last treatment unit).

There are also several one-tank systems. The most obvious example is a sequential batch reactor. In a batch system, the reaction environment varies for each stage of the treatment (fill, react, settle, and decant), as described by Bernardes and Klapwijk (1996). Continuous, single-tank systems with varied aeration/mixing patterns can also be designed. Even conventional biological wastewater treatment systems may have multiple environmental zones as a result of uneven mixing and aeration, which can result in nitrification and denitrification. This inherent capacity may be capitalised upon by a strategic wastewater-loading strategy. For onsite-generated wastewater, the flow is inherently variable. Onsite wastewater is generated in slugs resulting from specific water usage activities. This circumstance makes controlling the hydraulic loading very simple. The requirements are simply the addition of an equalisation tank (which can be a septic tank) and a timed dosing system.

The research reported here examined the possibility of maximising nitrogen removal from onsite-generated wastewater by carefully controlling the hydraulic loading (and thus organic and nutrient loading). The existing onsite system selected for the research uses 30 filter socks made from polyester felted material for solid/liquid separation. The socks also allow for the accumulation of a surface biofilm that may aid in denitrification.

Experimental Plan

The research was conducted in phases with the goal of maximising total nitrogen removal. In each phase, a variable or a set of related variables was altered, and the system operated until predictable results were obtained.

Experimentation Location

The research was conducted at the Western Regional Wastewater Pretreatment Facility in Montgomery County, Ohio. In 2001, the total average wastewater flow was approximately 13.5 million gallons per day (mgd), less than 10 percent of which was from significant industrial users. The wastewater for this research was obtained from the wet well of an intermediate lift station after use of coarse bar racks, use of fine screens, and gravity removal of grit.

System Layout

A commercially available onsite wastewater treatment system was used for the research. The unit had a capacity of 500 gal/day and was designed for a single-family residence. Solid/liquid separation was achieved with 30 fabric filter socks. There was a total of 135 ft^2 of filter surface area.

Influent was fed from a 500-gal/day circular tank that received wastewater from the Montgomery County plant, as previously described. The feed tank filled as needed, signalled by a level switch after a drop in volume of approximately 15 percent. A sump pump evacuated approximately 100 gal from the feed tank every 20 minutes to cause mixing and minimise excessive surface scum build up. The system was dosed in accordance with a set schedule that varied in each research phase. The on/off cycle roughly followed the American National Standard/National Sanitation Foundation Method 40 on

Residential Wastewater Treatment Systems. This schedule had three loading periods: morning, midday, and evening—6 a.m. to 10 a.m., 11:30 a.m. to 3:30 p.m., and 5 p.m. to 9:30 p.m., respectively. Some minor variations of this schedule did occur between phases. During each of the loading periods, the wastewater was fed for 1 to 6 minutes every 20 minutes. No wastewater entered the system at night (approximately 9:30 p.m. to 6 a.m.).

Analysis

Grab samples were collected directly in their storage bottles by graduate and undergraduate students. Sulfuric acid was added beforehand to preserve the samples to be analysed for nitrogen. All samples were transported to the laboratory in a cooler. BO[D5] and TSS analyses were conducted within four hours of collection. The other samples were stored in a refrigerator. Influent samples were obtained from the feed tank's recirculation line (used to keep the tank mixed, as previously described). Effluent samples were collected from the effluent line, which was continuously free flowing.

Hach Company methods and a micro-processor-controlled spectrophotometre were used for the nitrogen analyses and are believed to be adequate for research of this type, although they are not approved by U.S. EPA. Dissolved oxygen and temperature were not measured because of the difficulty of routinely and safely accessing the aeration portion of the treatment system.

Oxygen deficiencies were not expected, however, for the onsite wastewater treatment system used in this research. The wastewater temperature was expected to stay fairly consistent since the source was a publicly owned treatment works with a large flow (over 13 mgd) conveyed by a deep sewer system, and the onsite unit was in the screening room located at the same depth as the plant influent sewer.

To assess analytical quality, the authors analysed over 30 percent random duplicates. When a high-percentage recovery resulted, corrective action was taken, and samples in the suspect batch were rerun if possible; the data were discarded if the samples could not be rerun. Periodic blanks, standards, or both were also assessed for TSS, ammonia, nitrate, pH, and alkalinity. Results indicated that there were no quality problems (data not shown).

Throughout the research period, the onsite wastewater treatment unit demonstrated consistent and predictable operation with few upsets. The filter socks did not clog and were replaced only once between phases 3 and 4 as part of routine maintenance. Total nitrogen was calculated by addition of TKN (which includes ammonia) and nitrate, and the result is shown. Nitrite was not measured because it was assumed to be too low to influence the total nitrogen values. In "potential" total nitrogen represents the amount that would be in the effluent if no nitrogen removal occurred. The highest value during the entire research project was 15 mg/L; however, most values were substantially lower. Similarly, the effluent TSS concentrations were consistently low, with the exception of a few spikes that could not be traced to specific causes.

After the system had been operating for approximately 1 month; the effluent TKN was consistently below 10 mg/L of nitrogen (10 mg/L-N), regardless of the varied operating conditions. Similarly, ammonia effluent levels were, with some exceptions during phases 3 and 4, below 5 mg/L-N. The effluent nitrate tended to be slightly lower in the latter two phases. The level of total nitrogen in the effluent was lower, and more consistent removal of total nitrogen occurred, in the latter phases, when the instantaneous flow rate was substantially reduced. The daily average flow appeared not to be as important as the instantaneous loadings in reducing the total nitrogen level.

Nitrogen can be removed from wastewater by the following mechanisms:

- ammonia stripping;
- removal of solids containing nitrogen in the effluent TSS, wasted sludge, or both; or
- microbiological transformation of TKN and ammonia to nitrate followed by biodegradation of nitrate to nitrogen as a result of denitrification.

Ammonia stripping was not likely because at a close to neutral pH and without an engineered stripping unit, ammonia is largely in the ion form, which is not volatile. The only solids that were routinely released from the system were effluent TSS. These values were consistently very low. These results indicate that substantial microbiological denitrification occurred.

Another indicator of microbiological nitrification and denitrification is changes in alkalinity and pH between the influent and effluent. Alkalinity levels are reduced during nitrification, and while they increase by about half the amount lost during denitrification, the end result is a net decrease in alkalinity. Nitrification results in a decrease in pH, and denitrification results in a substantial increase in pH. The pH of the influent wastewater was lower in the last two phases of the research. As real wastewater was being used, this result can only be attributed to a change in user discharge characteristics. At a pH below 6.8, the rate of nitrification substantially decreases. There was also a corresponding decrease in ammonia removal. This result indicates a disproportionate amount of denitrification (compared with nitrification), which is not as sensitive to low pH values as nitrification. This trend is seen, which show substantial increases in pH in phases 3 and 4. Although the changing influent wastewater conditions prevent precise modelling, the above analysis provides strong evidence that biological nutrient removal occurred.

Enhancement of total nitrogen removal may be possible in any aerobic system through control of hydraulic loadings, the dissolved oxygen level, and mixing patterns. This research demonstrated that reducing the instantaneous influent flow by spreading it into periods of no flow allowed the loading of nitrogen to more closely balance the inherent nitrogen removal capacity of the specific unit tested.

The intermittent nature of onsite wastewater generation conveniently allows for this hydraulic-loading strategy. The only requirements are a storage tank for equalisation, a dosing pump, a controller, and a passive bypass in case pump failure occurs. It is not uncommon to find large septic tanks that can provide the basin for equalisation positioned before onsite wastewater treatment systems, which further simplifies the requirements for existing units. Each onsite wastewater treatment unit has a unique denitrification capacity that can only be determined experimentally. The onsite wastewater treatment unit used in the research reported here is believed to be particularly suited for high nitrogen removal because of the possibility of denitrifying bacteria attachment to the solid/liquid sock separators.

8

Water Supply and Wastewater Disposal

Providing an adequate amount of safe drinking water is one of the most fundamental responsibilities of a government, yet it cannot be assumed that all countries or local administrations have achieved this goal. The task is influenced by the quantity and quality of water at its source and by the technology available to process it. In communities relying on surface-water intake, the goal is closely related to the condition of discharged wastewater that becomes part of the freshwater resource base. This water quality frequently depends on the amount of available public finance and on the desire of government to install or purchase water-treatment facilities that will upgrade the water to the desired standard. In a broad sense, the situation may be viewed as a policy decision in which there is a trade-off among economic development, environmental concerns, and budgetary decisions.

The People's Republic of China must contend with water shortages as it tries to attain an adequate supply of safe drinking water for its burgeoning populace. As a result of the eighth five-year plan (1991-1995), it is expected that good water quality will become a national goal in the near future; yet there is little indication that the central government will make environmental spending a priority. The challenge to achieve good water has been met with limited success. Urban areas have become focal points of great concern because of their large, rapidly growing populations and because of the interest of the government in attracting foreign investment. Shanghai, the largest city in China, is the case study for examining a municipal effort to provide safe drinking water; the problems are associated with wastewater discharge, water intake, and treatment facilities, as well as governmental efforts to improve the overall quality of water available to the residents of Shanghai. The extent of the concern is exemplified by parents who teach their children from infancy not to drink water directly from faucets. All water for human consumption is boiled. The perception of dangerous water is well founded, because many illnesses and diseases can be linked directly to poor water quality.

Shanghai is a global-scale metropolis with a population of 12.5 million people and a position of prominence in commercial activity and industrial output in China. The old city forms the core of an area known administratively as Shanghai Proper, which is situated primarily on the western bank of the Huangpu River. The Huangpu drains an area of approximately 24,000 square kilometres and has a volume that exceeds 10 billion cubic metres annually. This channelled tributary discharges into the Yangtze River near its mouth as it flows into the East China Sea. Shanghai Proper is a densely constructed area of residential and other urban uses that must utilise an aging infrastructure; yet it is considered the most attractive part of the city. Surrounding the inner city are ten counties that provide high-value agricultural commodities. Portions of these counties are rapidly developing urban concentrations that will place increasing demands on water resources.

Freshwater Supply

Shanghai has always relied for its water supply on the Huangpu and its main tributary, Suzhou Creek, as well as on some canals. Although the Huangpu River will continue to provide virtually all of the water used in Shanghai, the quality of water from this source is usually below desired standards. Some water used to be pumped from wells as a supplement to the river water, but the practice was discontinued in the mid-1960s because it contributed directly to subsidence of land and related landuse problems. The annual natural recharge of groundwater was inadequate to maintain the historical hydrologic balance. Hydrologists measured the extent of this sinking land to be as much as 2.63 metres during the 1921-1965 period. The government attempted to solve that problem by pumping surface water back into the underground aquifers, yet during the mid-1980s the city continued to sink at an average of three millimetres a year. Because surface-water quality was often below acceptable standards, the procedure caused a decline in the quality of groundwater. A consequence is limited governmental options for future water supply because it is extremely difficult to cleanse a contaminated aquifer.

Before 1987 the twelve water-intake treatment plants in Shanghai were governed by the Shanghai Water Treatment Company. Eleven of these plants process water from the Huangpu River. The plant at Taopu is inland and treats water for industrial purposes only (Rei and Xia 1992). When the recently initiated project to remove water from the Yangtze was implemented, additional water-treatment plants became operational, and others are scheduled to become active before the end of the decade. In general, the water from the upper Huangpu is of much better quality than is the water that passes through the city in the middle and downstream sections of the river. The quality of water in Suzhou Creek where it joins the Huangpu is especially bad: it is highly toxic and has minimal oxygen available, with bubbles of gases along its entire course. An examination of seven water-treatment plants on dark and odorous days as measured by governmental standards reveals that the one at Yangpu, located at the confluence of the river and the creek, produces the most distasteful water. Water quality begins to improve downstream, where the Huangpu empties into the Yangtze, a pattern attributed to dilution by the relatively good water in the Yangtze that injects cleaner water back into the Huangpu twice daily as a result of tidal activity.

Sources of Wastewater Discharge

The municipal government has reported that Shanghai produces 5.4 million cubic metres of liquid waste daily. This figure does not include urban agriculture, urban leachate, shipping sources, or the surrounding area. The industrial sector contributes the largest portion of waste from Shanghai, and the National Environmental Protection Agency acknowledges industry as the main source of the pollution in China. Most industries and other intensive uses in Shanghai are situated along the Huangpu River. Currently, there are 12,400 industries in Shanghai Proper and an equal number in the surrounding counties. The dispersal of industry was designed to improve rural economies, to slow emigration to the cities, and to dilute the concentration of industrial pollution. All together the industrial sector produces 4 million cubic metres of industrial sewage daily.

Of the total industrial sewage, 58 percent is discharged directly into the Huangpu and its tributaries. The balance flows through a series of sewer lines and open ditches into the river system. Of this 42 percent, 70 percent eventually goes into the Huangpu and its tributaries, and the remainder is dumped into the mouth of the Yangtze River. This direct discharge into nearby offshore water was linked to illnesses that affected thousands of local residents who consumed contaminated seafood.

Among the most environmentally notorious industries in Shanghai are steel, chemicals, machinery, electronic components, textiles, and energy, which contribute about two-thirds of the industrial sewage discharge. Part of this discharge is toxic and may occur in significant quantities, so that it constitutes considerable public hazard. Countrywide during 1991, chemical industries accounted for 20 percent of the waste discharge and were the largest emitters of toxic substances that included mercury, phenol, chromium, cyanogen, or ammonium nitrate. In addition, other types of the industrial waste have been documented to have significant effects on oxygen deprivation in the water;

oxygen is essential to maintain desirable water chemistry. Yet another concern is the effect of the eleven coal-fired generating plants along the Huangpu that use very large quantities of water as a coolant and contribute to air pollution by burning sulfurous coal. The water is returned to the river system with only minor losses, although at the point of discharge the industrial water increases the ambient water temperature, which affects the biochemical processes occurring in the environment around the discharge points.

The balance of Shanghai's liquid waste, 1.4 million cubic metres daily, is attributed to residential uses. Over a large portion of Shanghai Proper the failure to install sewers during the early years of the city's growth substantially exacerbates the water-quality problem. Under these circumstances human liquid waste and gray wastewater become part of the storm-water runoff that flows along pavements and gutters. As it moves toward base level, it carries wastes and debris such as detergents, oil, paint chips, battery acid, cleaners, and solids. This untreated liquid empties directly into the Huangpu River system. Additional waste from some nonresidential inner-city landuses also contributes to the problem.

In recently built parts of the city, sewers transport some liquid waste to primary and secondary treatment plants that upgrade the quality of the waste before it is discharged into the river. Among the 1.4 million cubic metres of daily residential waste, only 0.21 million cubic metres are processed in treatment plants. The balance is collected in holding ponds behind residential complexes and then transported to the countryside or sent unaltered into streams and canals. Another source of untreated liquid waste comes from landfill leachate. This waste enters surface-water canals or aquifers and degrades local water quality. Depending on the composition of the waste material, toxic elements may be in the leachate. There are approximately 2,000 small urban landfills in the Shanghai metropolitan area. They have not been designed or constructed in accord with environmental standards defined by the federal government for the United States. These shallow landfills in alluvial deposits allow leachate to seep downward and laterally and to contaminate water that may ultimately become part of the intake water for Shanghai. Similar conditions are present at the three main landfill sites used by the Shanghai government.

In addition to urban pollution, a chief nonpoint source affecting water quality is the rural hinterland that is part of the watershed of the Huangpu River. China is one of the world's leading producers of chemical fertilizers, and agriculturalists are using increasingly large quantities of them, especially nitrogen, which are often applied in excess of crops' absorbency. The chemical overload, which includes pesticides and herbicides, becomes part of the surface-water runoff that seeps into groundwater and may prove harmful if people consume water directly from shallow wells. It is estimated that thousands of tons of nitrogen and phosphorus become part of the water chemistry of the Huangpu every year. An ironic consequence of the shift to chemical fertilization is the necessity of disposing of the principal component of night soil, which formerly had been the main source of organic fertilizer. In addition to human waste, pollutants attributed to livestock cause an oxygen deficiency in the Huangpu River.

Other harmful liquid waste is discharged from ships and small boats that use the port facilities on the Huangpu. Many vessels do not have holding tanks, and thousands of these craft move in and out of the port every day or remain at piers. The solid and liquid sewage produced daily amounts to 25,000 cubic metres. Also associated with waterborne traffic are occasional accidental spills and leakage of liquid cargo that could negatively affect water quality.

Wastewater Treatment

Wastewater treatment began in 1920, but a citywide system has not yet been completed. Over the years twenty-three primary and secondary wastewater-treatment plants have been constructed in Shanghai and the nearby countryside. These are grossly inadequate, because less than 5 percent of Shanghai's wastewater undergoes primary treatment. Many plants are located in densely populated areas, and the odors emitted during processing are frequently offensive to nearby apartment dwellers. Although heavy material is separated inexpensively, primary treatment does not include chemical treatment of the liquid. Therefore, large amounts of harmful liquid waste may be discharged into sources of domestic consumption.

Figure: *Wastewater Treatment Plant.*

The two forms of secondary treatment that are used rely primarily on activated sludge and, to a lesser extent, on biological membranes. The activated sludge is a cultivated biological material with microorganisms that blend with wastewater to absorb and decompose certain unwanted components. In a somewhat similar manner, a biological membrane composed of microorganisms and organic and inorganic compounds performs the same function but is faster than is activated sludge. Both of these treatment methods are common in the United States.

Some of the twenty-three wastewater-treatment plants are operating above design capacity; several have not attained their designed efficiency rate; and others are in disrepair and need to be renovated. Fifteen plants treating residential and some industrial waste are located in the city and have a design capacity of 490,000 cubic metres daily. Only nine of them were operating in the early 1990s, when total wastewater treatment amounted to 180,000 cubic metres daily. Eight treatment plants are located in the nearby countryside. Designed to treat 87,000 cubic metres daily, they are processing only 30,000 cubic metres daily. Plants that are operating above capacity are not treating the water efficiently, which therefore leaves the process only slightly more effective than dumping the excess untreated water directly into the river system.

Figure: *Wastewater treatment plants*

Industrial plant managers in China are responsible for treating toxic and less harmful waste produced at their factories. The typical process focuses on the use of organic or inorganic cohesive material to condense pollutants in sewage and to make them easy to separate. This type of treatment is designed to remove toxic pollutants in the water, such as some heavy metals and phosphorus and other harmful waste products from factories devoted to pulp and paper making, food processing, steel and chemical production, cloth dying, and mining operations. Other methods include the use of active carbon, static electricity-ion exchange, membrane separation, and ozone oxidation. Unfortunately, very few factories have proper treatment facilities for toxic materials, and many are not used efficiently because operating them is very costly. The problem has been compounded by the historical development of wastewater-outfall and water-intake sites. According to China's drinking-water resource protection law, no discharge is allowed within 1,000 metres upstream of a water-intake station. The law also states that no intake station should be located within 1,000 metres in either the upstream or downstream direction of a tide-sensitive river. However, along the Huangpu, which is tide sensitive, an average of 4.5 industries within every 1,000 metres discharge wastewater near intake stations. Furthermore, some intake and discharge stations that were operational before laws regulating pollution were adopted are in very close proximity. Wastewater does not flow far enough to be diluted before it is taken in by water-treatment plants.

Although the government may monitor the waste-discharge process, water-quality standards may not be as strictly enforced as they are in countries with a significant commitment to the environment. When citations are issued in Shanghai, the fines for violations are so low that they do not induce managers to change their plants' discharge practices. Polluters think the monetary fine will be less than the long-term cost of installing or operating the equipment properly.

There is no policy that can force managers to operate the waste-treatment facilities to their maximum capability. The scope of the problem is exacerbated by newly opened small factories in outlying areas where treatment facilities seldom exist. Equipment in other factories may not be maintained well and may break down. Monitoring water quality is unsystematic and infrequent, largely because of the lack of an enforcement agency designated to monitor the discharge operations in the middle and lower portions of the Huangpu River.

Improving Water Quality

Two sewer lines were constructed in the 1970s to transfer untreated wastewater from the urban area. The southern line is a combination of pipeline and open ditch. The upper section, known as Rihuei Canal west of the Huangpu, has been converted to sewer lines, and portions of the unprotected ditch east of the river are continually being renovated by covering their surfaces to prevent accidents and people throwing waste material in the water.

Together with the western pipeline, they remove 550,000 cubic metres and 250,000 cubic metres of liquid daily respectively. Because the water is contaminated, the result is "dark belts" that have been measured as being 3 x 20 kilometres and 1.3 x 10 kilometres in an area where the two sewer outfalls empty into the Yangtze and the East China Sea. Although pollution in the Huangpu has been reduced, the procedure degrades offshore water, which is counter to a goal of the latest five-year plan to promote safe seafood for consumption. The Shanghai municipal government enlisted the cooperation of technical specialists from Australia, Great Britain, and the United States to prepare a report entitled "Huangpu River Pollution Integration Treatment Plans."

Figure: *Improving Puget Sound Water Quality through "Green Street"*

Approved by the central government in 1986, the report advocated measures to alter the characteristics of the Huangpu River so that the quality of the water-intake system would be improved. As a partial short-term solution, the upriver portion of the Huangpu was designated the primary source of water, and the middle and lower portions of the river were assigned to transportation and sewage discharge. The sites of water intake and outfall were to be placed farther apart. Two additional recommendations were the development of a river-water diversion project and the creation of an upper Huangpu River preservation policy for long-term protection of the headwaters. The government has established standards for acceptable water intake, with the goal of reaching national surface-water-standard class two or three on a scale of one to five for the entire Huangpu. In 1986 only 24 percent of the total flow of the Huangpu contained water exceeding class three, which is within the acceptable limit of drinking water.

Shanghai Sewage Project

The largest and most meaningful wastewater project in recent Chinese history centres on improving the quality of Suzhou Creek. This project is the first stage of the Shanghai Sewage Project. As a follow-up to the 1980 plans, an agreement was reached in May 1983 between the Australian and Chinese governments and the World Bank to proceed with a detailed sewage management-strategy study. After June 1984 seventy Chinese engineers and specialists further developed a preliminary design for a third supersewer that would augment the southern and western pipelines.

Figure: *Shanghai Sewage Treatment Project*

The Shanghai Municipal Engineering Administration Bureau was assigned the responsibility of supervision. A team of consultants from the Australian Development Assistance Bureau would support this agency with technical information when requested. In October 1986 the design was formally approved, and construction began in August 1988.

Known as the Eastern Line, this pipeline extends more than 30 kilometres from west of the Huangpu into Chuansha County and serves 2.5 million people. Shanghai scientists are certain this project will improve water quality in both the upriver and downriver portions of the Huangpu.

Designed to discharge 1.4 million cubic metres daily, the project collects sewage from factories and residential quarters around Suzhou Creek that is diverted to the mouth of Yangtze River. The main gravity interceptor was completed by the end of 1993. In addition to it, fifty-three pumping stations are to facilitate waste movement. Near the point of discharge a pretreatment plant removes objectionable floating materials and solids that worsen contamination of the Yangtze River.

Chinese wastewater-management specialists concluded that because actual discharge would occur 1.5 kilometres from shore and at a depth of fifteen metres, the sewage would be diluted within one to six kilometres. They also asserted that the detrimental impact would be negligible. Although the assertion may be initially valid, there must be concern about the long-term result of this approach to processing wastewater.

Now that the main sewer line is operational, a second stage of the project designed to collect and discharge sewage from extended parts of the city adjacent to Suzhou Creek has been initiated. The Chinese expect that this engineering project will further improve the quality of Huangpu water.

Zhonggang Discharge Sewage Project

In May 1986 planners began to discuss a fourth large sewer pipeline that would extend from the Hongqiao area near Suzhou Creek across the upper Huangpu River to Zhonggang on Hangzhou Bay. This line would extend seventy kilometres and require seven or eight pumping stations. The Zhonggang Sewage Project was designed to collect sewage from Wujin and Minhang industrial areas and other wastewater producers. The result would be substantially improved water quality in areas around the Linjiang and Daqiao water-treatment plants. The design capacity was approximately 1 million cubic metres daily and would be devoted primarily to transporting industrial wastewater. Although the benefit of the project was clearly evident, it was cancelled due to budgetary constraints. Now, however, it is being reconsidered.

Upper Huangpu River Diversion Project

In 1983 Chinese water-resource specialists approved the concept of a two-part upper Huangpu River diversion project. The project began in February 1985. The initial construction, known as the Pudong Part, consisted of a series of pipelines and canals from Linjiang to five downstream water-treatment plants. This portion allowed the treatment of relatively good water instead of polluted water from the middle and lower courses of the river. Pudong Part was completed in July 1987.

The Linjiang site was initially designed only as a pumping station to transport untreated water from the upper Huangpu directly to downstream treatment plants. Included in the early plans was a project to pump water through a pipeline from Daqiao to Linjiang, which would allow the river water to bypass the industrial concentration around Minhang and Wujin. The pipeline would improve the water quality at Linjiang on most occasions, yet tidal action on the Huangpu River would cause the water-treatment plant at Daqiao to contend with some polluted water backing upstream from the main industrial areas. Unfortunately, funding was not available at the time to complete the work necessary to build the long Daqiao section of the diversion pipeline, but capital was sufficient to upgrade the planned Linjiang pump station to serve as a water-intake station and water-treatment plant with a daily capacity of 2.3 million cubic metres.

A 1988-1989 survey after the shortened plan for the diversion project assessed the quality of river water near Linjiang and Yangpu. Measurements indicated that average nitrogen levels and available oxygen counts were better at Linjian than at Yangpu, but they were still not satisfactory. Although water quality at Linjiang generally has a higher national water-class standard, sometimes the quality of the two sites is reversed. The varied water quality at the Linjiang water-treatment plant is attributed to the tide sensitivity at this point in the Huangpu. Here polluted water from the middle and lower river retreats upstream and affects water quality. Also along this stretch are seventeen large industries within 3.5 kilometres of the Linjiang treatment plant. Somewhat farther upstream are main industrial areas such as Wujin and Minhang, where the water is polluted badly by industrial discharges.

A second construction activity, part of the upper-river project, began in 1987: revival of the initial plan to relocate main water-intake stations to the headwaters of the Huangpu River. This portion of the project will require a return to the original plan to build an enlarged new water-intake station and water-treatment plant at Daqiao. The plant will replace four intake stations along the lower river and one in its middle section. The Daqiao plant is upstream from all main industries and withdraws water of class-two or class-three standard. It will have an intake capability of 5 million cubic metres daily. This phase of the project is expected to be completed by 1995.

Upper Huangpu River Preservation Plan

Only the upper portion of the Huangpu River beyond Longhuagang occasionally achieves the class-two standard, because point and nonpoint pollutants enter the drainage system from adjacent provinces. Dianshan Lake, near the headwaters of the Huangpu, contains the best-quality water, but this clean and beautiful environment is becoming an important tourist attraction. Retirement and convalescent homes, sports-training facilities, a large hotel, and related facilities have been built around the lake. The attractiveness of the site has led to intensive use and to deteriorating water quality. Waste discharged from nearby developments has caused some eutrophication of the lake. At various times of the year a large portion of the lake turns green.

Because one cornerstone of providing safe water is to withdraw it near the sources of the Huangpu, it is necessary to protect this water quality. More than a million people reside in the six counties and three districts that constitute this 830-square-kilometre headwater area. More than 300 point and nonpoint sources of pollution have been identified in the region, including twelve large factories that discharge 85 percent of the total sewage in the Upper Huangpu. Most of the factories are chemical plants or paper mills.

Governmental policy toward these factories is based on a calculation that allows a certain amount of pollution from each factory. An official notifies the factory of this value and then grants permission to begin operation. Each factory-management team is required to build and operate a wastewater-treatment facility and to adjust its manufacturing process to reduce the wastewater discharge. The type and amount of waste discharge must be certified, and law enforcement must become more diligent. Similar waste-discharge plans are scheduled to be required for small rural factories. Factories that were difficult to upgrade were threatened with closure, and new factories would be properly designed to reduce discharge and improve wastewater quality. Shanghai also began to address rural residential waste and nonpoint sources from agricultural activities. More efforts were planned to collect and treat water properly in areas of intensifying development. One of the agricultural initiatives is to use more animal waste for fertilizer and to build a methane-gas collection system. Officials also recognise the need to control the quality of water that enters Dianshan Lake and the drainage system from Jiangsu and Zhejiang provinces.

Future Water Supply

It is estimated that by 2000 Shanghai will require 7 million cubic metres of water daily. Approximately 70 percent of this total will come from the Upper Huangpu region; the other 30 percent must be obtained from other sources. Chinese technical experts decided to withdraw the necessary balance from the virtually unlimited supply in the Yangtze River. There is some concern about this decision, because the river is subject to tidal effects. Seawater and freshwater meet between the mouths of the Yangtze and Huangpu rivers. In the dry season and at high tide, seawater can extend beyond the discharge point of the Huangpu. An additional concern is that during its long flow to the ocean the Yangtze receives influxes of low-quality wastewater. The best-quality water is located between Baogung Reservoir and Xuliujin, 45 kilometres west of the Baogung steel plant along a southern bend of the Yangtze River. Most of that portion of the river is in Jiangsu Province. It is predicted that the water quality there will be maintained at national class-one and class-two standards by 2000.

The diversion project to withdraw water from the Yangtze was initiated during 1993 and 1994. Two water-intake plants at Wusong and Zabei, devoted to removing water from the Huangpu River, were refitted as treatment plants for water coming from the Yangtze. In addition, two new plants at Yuepu and Linqiao began to treat Yangtze water. The total capacity of these plants remains very low, but the government is committed to their enlargement.

Furthermore, a plant at Dachang was scheduled for construction during mid-1994, with completion expected by the end of 1996. Another plan involves diversion of water from the Yangtze River at the city of Wuhu, upstream from Shanghai. The water would be sent a long distance to Taihu Lake, west of Dianshan, an area that reportedly needs about 80 percent more water than is locally available to provide for local irrigation, to maintain adequate water flow, and to prevent polluted water from reaching intake plants on the Huangpu. The flow from the headwaters of the Huangpu is reduced in summer because a considerable amount is used for agricultural irrigation. Reduction of the water volume usually causes the polluted water to flow farther back upstream toward the lake area. Diversion at Wuhu would supplement the water supply during periods of low flow.

At the beginning of 1990 the Shanghai Water Conservancy Bureau considered constructing a reservoir at Qinchaosha on the northern edge of Changxin Island. The Qinchaosha area is growing steadily, and water quality there is fairly high, because there is little industrial pollution on Changxin Island. A reservoir there could meet the 2000 water-supply and water-quality requirements of the new economic development zone east of the Huangpu. A reservoir would be desirable to supplement water from the existing sources. It would be filled during low tide and would be closed during high tide to avoid saltwater contamination. The reservoir could supply 2.5 million cubic metres of water daily during the dry season, and its storage capacity could be expanded.

The water quality in Shanghai is a principal environmental concern of officials and residents; yet actions to remediate the problem have not completely resolved the issue, even though activities are continually occurring to achieve higher standards for drinking water. Currently a substantial portion of water from the Huangpu River is below national surface-water standards, because wastewater-treatment facilities are inadequate. These facilities are too few in number, are in disrepair or are overloaded, and are not designed or constructed to process the exceedingly large quantity of toxic industrial waste. Other sources of harmful water include wastewater from unsewered urban developments, minimally designed landfills, waterborne craft, and rural landuses.

A goal of the eighth five-year plan is improved water quality. Two efforts to achieve this goal are construction of large sewer pipelines to divert poorly treated wastewater to the mouth of the Yangtze River or to the East China Sea and an upper-river diversion-preservation project at Daqiao, which is generally removed from the effects of large industrial complexes. Although the source of good-quality water is projected to be satisfactory, the amount of available water is reduced somewhat, while the need for water has increased. A water-resource plan is being implemented to withdraw more water from the Yangtze. As the quality of intake water improves in the source area, concern remains about upgrading wastewater-treatment facilities. Models have been developed to indicate that unprocessed water will not cause long-term degradation of the environment, but they may prove unreliable because of the unprecedented quantity of untreated water that is being emptied into nearby water bodies. Part of the solution requires greater governmental commitment to improve wastewater treatment by factory managers without reducing industrial profits. This in turn will necessitate a larger allocation of capital to upgrade municipal treatment facilities along the Huangpu River.

Wastewater Management - Approach to Improving Water Quality and Preventing Disease

The postwar economic and population growth of the 1950s launched many families into suburbia in Hamilton County, which is Ohio's most southwestern county and location of Cincinnati. The expansion of Hamilton County followed a typical pattern: Construction of suburban residential housing outstripped the development of traditional urban infrastructure. Extended aeration technology made it possible to put a miniature wastewater treatment plant in the yard of every new homeowner. Aeration systems were marketed as virtually maintenance-free and capable of producing effluent with the quality of drinking water. These misrepresentations, in concert with the natural limitations of the sites being developed and a willingness in the community to believe something too good to be true, brought public health consequences.

Since most of Hamilton County escaped the advance of the last glacier, steep shale ridges remain covered with thin top soils and slowly permeable silty-clay subsoils. Only small areas in the valleys of the three principal streams are underlain by the more permeable glacial deposits. Seasonal rain and snow-melt runoff cut the ridges with innumerable erosion channels that join to form intermittent streams. Conditions that limited the use of leaching devices appeared, during the Hamilton County expansion, to be optimal for aeration systems with surface discharges. Swayed by exaggerated claims of operational efficiency, public officials were quick to accept the technology as a way to support new tax-generating development without the cost of sewer construction. Developers enthusiastically supported a public policy that reduced their capital costs and raised profits. Homeowners believed that they had effective, low-cost, state-of-the-art sewage treatment devices. With discharges running to any downhill drainage way, out-of-sight-out-of-mind became the prevailing attitude. It took a long time to change. By the mid-1980s, sewage contamination of the West Fork of Mill Creek elicited strong public reaction. When a hepatitis A outbreak occurred in 1989, the public clamour over lack of public sewers, ineffective sewer system operation, and poorly functioning aeration systems intensified. Public alarm further increased when a child playing in a stream polluted by aeration systems caught a rare protozoan infection and was hospitalised for days.

Table : *Home Aeration Units - Percentages by Manufacturer*

Manufacturer	*Percentage*
Cavitette	22.4
Coate	10.6
Jet	36.6
Multiflo	2.4
Norweco	0.1
Oldham	27.9

By the early 1990s, the total number of aeration systems was unknown, but estimates ranged from 20,000 to 40,000. Public perception regarding the public health impact was markedly changed. An irate homeowner persisted until the Board of Health of the Hamilton County General Health District officially declared her neighbourhood a public health nuisance and requested the state to order construction of a public sewer.

Public concern peaked when the Board of Health authorised use of aeration systems in a proposed subdivision from which effluent would drain directly into the recreation lake of homeowners downstream. The resulting legal action brought sanctions from the Ohio Environmental Protection Agency (OEPA) and the Ohio Department of Health against the Board of Health. A political response occurred in 1993; board members were replaced and a new health commissioner (a public health officer) was employed. The new board determined that water quality improvements and waterborne disease prevention were dependent on the effective use of household sewage disposal systems. Thus, the board made a commitment to:

- update its household sewage regulation, which had been in force since 1959;
- reconsider its previous approval actions regarding the use of aeration systems in residential subdivisions;
- inventory and evaluate existing home sewage systems;
- build community partnerships and educate homeowners about system operation and maintenance requirements;
- expedite investigations of complaints; and
- strengthen enforcement when sewage system repairs are necessary.

This chapter discusses an operation permit program that was established to fulfill several of these commitments. The components of the program are described, as are the steps that have been taken to improve the water quality of Hamilton County streams.

Building the Operation Permit Program

Policy Development

In Ohio, local health districts have sewage system permitting authority for one-, two-, and three-family dwellings. These districts enforce either the minimum state code or more stringent regulations adopted by the local board of health. Under the minimum code, a newly constructed system is automatically permitted for operation upon final construction approval. This operation permit remains in force until it is revoked by board action. Such permits are rarely revoked. Some health districts issue operation permits for existing sewage systems. In 1994, the Hamilton County General Health District (HCGHD) initiated a comprehensive operation permit program. Under this program, operating permits are issued both for new and for existing household sewage disposal systems.

On December 13, 1993, the Hamilton County Board of Health adopted a new household sewage disposal code and created a new division to enforce it. Previously, the sewage-permitting functions were a part of the health district's plumbing division. To implement the board's commitments, the Division of Water Quality and Waste Management was established. The main function of this division was to implement Regulation 529, the Household Sewage Disposal Code.

To guide the new program, the Board of Health established the following policies:

- No permit fee will be billed until an inspection of the household sewage disposal system (HSDS) is completed and the owner is provided with a written inspection report.
- All HSDSs with electrical components (aeration units, etc.) are subject to an annual inspection. If a system is found to be operating properly, a one-year operation permit will be issued.
- All nonmechanical or nonelectrical HSDSs will be inspected once every three years. If an HSDS is found to be operating, a three-year operation permit will be issued (effective September 11, 1995).
- All newly constructed dwellings with newly installed HSDSs permitted and approved by HCGHD will be exempt from the operation permit program for a period of two years following construction approval.

Critical to the success of the operation permit program was the establishment of inspection criteria for determining proper operation. Because of the varying ages of the systems in use and the different standards under which they were manufactured and installed, a balanced approach was needed that would achieve water quality objectives while maintaining public and political support. The health commissioner and staff were charged with the development of the inspection criteria.

It was decided that system performance would be evaluated by observation rather than effluent sampling. This decision took into account the fact that many HSDSs have no separate access for sampling, as well as the probability that most older systems would not be able to meet effluent quality standards established by the Board of Health. Also considered were the cost of sampling as many as 20,000 sewage systems and the fact that many individual systems were connected to common collector lines. (Collector lines function as private sewers that transport effluent from several homes). Ten observational criteria therefore were used to establish improper performance of a mechanical HSDS, leading to the designation "disapproved":

1. motor missing;
2. motor inoperable (cold);
3. motor not drawing air or insufficiently drawing air;
4. broken lid (i.e., piece missing or broken to the extent that it allows entrance of surface water, or lid cannot be lifted without collapse, or metal grating is decayed);
5. flooded filter;
6. visual evidence of septic sewage (i.e., black, odorous);
7. visual evidence of electric service problem;
8. components not functioning in accordance with design standard;
9. discharge creating a public health nuisance; and
10. access riser has not been brought to grade over each compartment requiring maintenance.

Table: *Disapproved Systems with Completed Repairs*

	Number of Systems	***Number of Reinspections***	***Mean Reinspections to Compliance***
Totals	2,741	3,861	1.41

Policies concerning program fees have changed as program activities have evolved and knowledge has been gained about operating costs and the willingness of HSDS owners to pay. At the outset, the Board of Health established a single operation permit fee of $40. As program implementation progressed, two problems became apparent. First, there was no way to recover the cost of the additional inspections required to achieve and confirm system repairs when the initial inspection resulted in disapproval. Second, no permit was issued to HSDS owners who failed to pay their permit fee. Consequently, those who were delinquent soon forgot about the need for an operation permit as well as their indebtedness. HCGHD then entered a new era, one marked by the use of dunning letters and appearances in small claims court, and found itself acting somewhat like a small utility company.

In July 1995, the fee structure was changed on the basis of a report issued to the Board of Health: Aeration Sewage Disposal Systems in Hamilton County. This report evaluated the progress made to date. With community input, the board implemented a variable fee structure. The annual fee covering the initial inspection and permit was reduced to $30. A reinspection fee of $30 was established for second and subsequent reinspections. As an inducement for HSDS owners to provide regular maintenance, the second reinspection fee was set at $15 for owners holding a maintenance contract with a registered and bonded service company.

Delinquent permit fees rose as high as 62 percent but averaged around 13 percent annually during 1994-1995. By the end of 1995, uncollected permit fees amounted to $32,320. In 1996, HCGHD retained a collection agency at a cost of $4.95 per outstanding account. Also, a $10 late fee was assessed against each delinquent account. Delinquent fees dropped to an average of nine percent in subsequent years. Many homeowners felt more threatened by a bad credit rating than by criminal prosecution. Nevertheless, the percentage of unpaid accounts was still too high. In 1998, HCGHD worked with local Ohio legislators, and legislation was passed allowing health commissioners to certify unpaid operation permit fees to county auditors for placement as liens on property. With the passage of this legislation, the playing field has been levelled so that all HSDS owners have a financial stake in the operation permit program.

Staff Development

The development and implementation of policies regarding HCGHD staff and staff activity have been crucial to the success of the operation permit program. These policies have been established largely by management and include issues such as personnel qualifications, inspection protocol, and training.

Table: *Nonmechanical Sewage Systems 1996-1997*

Type	*Number*	*Type*	*Number*
Drywell	783	Drywell	162
Leach lines	370	Leach lines	476
Subsurface sand filter	181	Subsurface sand filter	1
Privy	1	Privy	212
Other	84	Other	27
Total	1,419	Total	878

Total initial failure 59 (4.2%) Total initial failure 117 (13.3%)

Field work was done initially by registered sanitarians and sanitarians-in-training. During the first year, staff turnover was unacceptably high. Newly graduated environmentalists quickly gained valuable field experience but were just as quickly tired by the large volume of inspections and the repetitive, sometimes confrontational nature of the work. Experienced sanitarians soon found the work not to be challenging. Management found a solution in the employment of technicians as inspectors. With specialised training and sanitarian oversight, these staff members have maintained their enthusiasm and perform their inspections in a highly competent and efficient manner. The water quality technicians have worked well with HSDS owners and the repair contractors, too. The effectiveness of the inspection staff can be attributed to the development of operating rules that emphasize thorough training, clear identification of staff, and consistency in performance of duties. Each inspector receives detailed training in the operation of each brand of aeration system that he or she will inspect in the field. Contact with repair contractors is encouraged. When possible, attendance at a manufacturer's training school is supported. Inspectors also receive training in communication and public relations, and membership in professional environmental associations is encouraged.

Inspectors adhere to a dress code, wearing distinctively coloured shirts and jackets and clearly visible photo identification badges. Each inspection begins with a stop at the front door, where the inspector explains the purpose of the visit to the citizen. Next comes a brand-specific standardised inspection of the sewage system. Owners who are present are informed of the inspection findings.

The Inventory of the Problem

Aeration systems contain electrical components and filters that require maintenance by homeowners. Inspectors found that homeowners were unaware, for the most part, of these maintenance needs. The media reported that 40,000 aeration systems were discharging untreated waste water into homeowners' backyards, although the exact number of such systems was unknown, as were their operating conditions. Installation permit records were not complete. While citizens worried about disease and pollution, an OEPA connection ban and class action lawsuit put the Hamilton County Board of Health in a reactive mode. There was a rush to inventory and evaluate the operating status of all existing aeration systems first. The Board of Health directed the health commissioner to address the following questions:

1. How many aeration systems exist in Hamilton County?
2. What are the manufactured types?
3. Where are the aeration systems located?
4. What is the operational status of these systems?
5. How much water pollution and how many health nuisances have been created from these aeration systems?

Table: *Pass/Fail Inspections of Home Aeration Sewage Systems*

Time	*Number*	*Percentage*	*Failed*
Period	Inspected	First	Inspection
1994-1995	14,992	33.1%	(4,962)
1996-1997	17,685	6%	(1,061)

From January 1994 through July 1995, sanitarians and water quality technicians blanketed the county, inspecting aeration systems. The staff visited over 10,000 properties - many on several occasions to find aeration systems, locate collector lines, consult with homeowners, or reinspect systems for compliance. Inspection sheets were filled out, and information was entered into a custom-designed database for ease of tracking. Within 18 months, 9,145 aeration systems were located and inspected. Six types of manufactured home aeration systems were found. A significant concern was the high percentage of Cavitette brand systems in use, because there was no active manufacturer and no replacement parts were available.

The Cavitette brand was manufactured locally from the late 1950s through the 1960s. The design of this system was based on less stringent standards than NSF Standard 40. Of the 9,145 aeration systems inspected, 34 percent, or 3,077, were disapproved between January 1994 and July 1995. Many of the aeration systems were connected to common yet private sewer lines known as collector lines. These collector lines served as few as two homes or as many as 50 homes. The effluent from the collector lines discharged into stormwater sewers, ephemeral streams, and onto the ground surface. None of the collector lines had National Pollutant Discharge Elimination System (NPDES) permits. It had been observed and reported that the discharges from these collector lines had degraded the water quality in the county's streams and waterways. Before 1994, there was no documented wastewater monitoring for these collector lines. In the summer of 1994, however, as a part of the operation permit program, a collector line evaluation and assessment of needs project (Project CLEAN) was implemented. Wastewater samples were taken from 197 collector lines. The samples were analysed for biochemical oxygen demand (BOD), suspended solids (SS), and fecal-coliform bacteria.

When the Board of Health adopted a new sewage code in late 1993, it also codified discharge standards for SS, BOD, and fecal-coliform bacteria, as follows:

- BOD - 20 mg/L;
- SS - 40 m/L; and
- fecal-coliform bacteria - 5,000 colony forming units (CFU) or less per 100 mL.

Nearly three-fourths (72 percent) failed to meet the standard for fecal-coliform bacteria. This was not surprising since disinfection devices had not been installed on many of the individual aeration systems. The initial inventory and evaluation of aeration systems was completed.

Nonmechanical Sewage Systems

A contention by County citizens early in the program was that HCGHD was ignoring the pollution created by nonmechanical household sewage disposal systems, like leach lines and dry wells. Once the initial inventory and evaluation of aeration systems was completed, HCGHD staff began the inventory and evaluation of nonmechanical sewage systems in 1996.

***Table** : Mean Values from Stream Sampling*

Sampling Date	*BOD*	*Fecal*	*N[H3]*	*SS*
3/12/97	1.63	435.98	0.07	7.34
6/19/97	6.26	1,1541.46	0.13	26.29
9/25/97	4.45	1,637.5	0.27	6.9
12/17/97	5.22	5,007.73	0.6	6.91
4/2/98	5.18	4,546.17	0.5	14.59
5/6/98	5.91	31992.93	0.54	113.69

Minimum n = 40

The inventory and evaluation of nonmechanical sewage systems is a work in progress. By May 17, 1998, a total of 2,605 nonmechanical sewage systems had been assessed. The following inspection criteria are used to approve or disapprove a nonmechanical household sewage disposal system. Approve if:

- system components can be located, the system uses in-soil dissipation, and there is no surface seepage of gray, malodorous effluent (minor seasonal wetness is acceptable); or
- the system uses surface discharge; the discharge pipe is accessible with sufficient freeboard to allow collection of an effluent sample; and the effluent is clear, not malodorous, and not ponding on the inspected property or on an adjacent property.

Disapprove if:

- gray, malodorous sewage is seeping to the ground surface, creating a nuisance;
- gray, malodorous sewage is discharging from the sewage system to an adjacent property or roadside drainage way; or
- any similar condition is occurring that creates a public health nuisance.

Upon the initial inspection, the percentage of nonmechanical systems disapproved overall averaged nine percent. The inventory and evaluation of these sewage systems continues. HCGHD did not meet its goal of inspecting and inventorying all nonmechanical sewage systems by December 31, 1998. On March 8, 1999, the Board of Health changed the nonmechanical permit from a three-year permit to a five-year permit.

Building Community Partnerships

All the media attention over the lack of governmental oversight and widespread pollution did not convince the vast majority of citizens with home sewage systems that the operation permit program was necessary. Media and citizenry criticism of HCGHD's historic lack of oversight was replaced by concern about HCGHD persistence in requiring sewage system repairs and payment of permit fees. During the first 18 months of the program, HCGHD staff handled 20 to 40 telephone calls per week. About 265 citizens felt strongly enough to write letters protesting the program.

The themes ranged from questioning the necessity of inspecting septic systems to complaints about government invasion of property rights and privacy HCGHD initiated a community education program to gain public acceptance of the operation permit program, to improve political support, and to teach homeowners about HSDS operation and maintenance. Brochures were developed and mailed to homeowners and handed out at neighbourhood festivals and community halls. Numerous press releases were sent out, and editorials were published in the local newspapers about the benefits of the operation permit program to the community. HCGHD staff conducted presentations at neighbourhood gatherings or backyard barbecues about the importance of HSDS maintenance.

The Metropolitan Sewer District (MSD) and HCGHD forged a partnership and collaborated at several public meetings. The two agencies worked in concert to extend public sewers into those watersheds where sewage nuisances were prevalent and HSDS upgrades were not feasible. A public sewer assessment credit was established by the Hamilton County Board of Commissioners. This credit was brainstormed by MSD and HCGHD officials together with a group of western Hamilton County business leaders and elected officials. The credit program made all single-family dwellings with HSDSs in existence on or prior to September 20, 1995, eligible for a $5,000 credit toward their public sewer assessment. The public sewer assessment credit helped convince many homeowners to sewer their neighbourhoods.

Using GIS

In 1996, HCGHD began using the Cincinnati Area Geographic Information System (CAGIS). The CAGIS technology allows HCGHD to place all home sewage systems, stream quality data, and communicable disease data on computer-generated maps. Different kinds of information can be overlaid for analyses; for instance, the public sewer system layer can be overlaid with the county home sewage system layer, allowing the user to quickly observe the proximity of public sewers to home sewage systems. Other layers of data, such as stream data and communicable-disease information, are overlaid to find patterns or clusters of disease or pollution associated with home sewage systems. Because it helps target resources, the CAGIS technology is a powerful public health surveillance tool.

Results - Has the Operation Permit Program Made a Difference? Data from two time periods, 1994-1995 and 1996-1997, was compared to reveal program successes and failures. As of January 1, 1998, HCGHD staff located a total of 9,515 aeration systems. The staff continues to find additional aeration systems in remote areas of the county. Comparing the data for the two time periods clearly shows a large reduction in the number of systems failing the first inspection. Homeowners are ensuring that their aeration systems are maintained. The number of homeowners with private maintenance contracts increased from 1,623 in 1995 to 2,274 in 1996. In 1997, however, the number of homeowners with private maintenance contracts decreased slightly to 2,220.

Additional sampling of collector lines was conducted to determine program effectiveness. A 12 percent subset was randomly selected from the original Project CLEAN sampling locations, and the locations in this subset (23 locations) were sampled and analysed for BOD, SS, and fecal-coliform bacteria. The collector line discharges were sampled once each year over the course of a three-year period. All samples taken in 1997 show a fourfold decrease in BOD compared with samples taken in the first two years of the program. Similar decreases in suspended solids and fecal-coliform bacteria occurred. These results demonstrate that the operation permit program for aeration systems connected to collector lines has reduced the potential for disease conditions and diminished pollution in the affected neighbourhoods. The stream sampling program was initiated in 1997 as an outgrowth of Project CLEAN. Forty sampling locations were selected. In 1997, quarterly stream samples were taken from these locations in an effort to observe differences in water quality over time. During sample collection, stream velocity, air temperature, and overall weather conditions were noted. The goal of the stream-monitoring program is to assess the water quality impact within the watershed. There are not, however, enough data points to establish any trends other than random variation.

Citizen Opinion Survey

During the spring of 1998, HCGHD mailed out 1,000 customer surveys to homeowners with aeration systems. The sample was selected by the number on the cash register receipt in the home aeration system database. The survey asked questions about program effectiveness, program personnel, and program components. The response has been good. As of June 5, 1998, 314 survey questionnaires were completed and returned.

In response to questions about program effectiveness, slightly more people agree than disagree that the operation permit program has reduced odors, has improved stream clarity, and provides a valuable service. In written comments, nine respondents wanted sewers, seven offered compliments, nine noted fewer odors, six wanted lower fees, five questioned the fairness of the program relative to pollution from semipublic sewage plants, three resented government intrusion, and 11 considered the operation permit program redundant because of private maintenance contracts with service companies.

***Table** : Citizen Survey Results(*)*

	Mean of Responses	***Yes***	***No***	***Standard Deviation***
Program Effectiveness				
Odors have improved	2.92	-	-	1. 1
Stream clarity improved	2.96	-	-	0.95
Visible sewage decreased	2.95	-	-	0.9
Service good value	2.83	-	-	1.18
Fees affordable	2.91	-	-	1.19
Program Personnel				
Inspectors identifiable	2.59	-	-	0.92
Inspectors courteous	2.17	-	-	0.76
Inspectors helpful	2.41	-	-	0.85
Office courteous	2.56	-	-	0.81
Office helpful	2.61	-	-	0.8
Program Components				
Information adequate	2.35	-	-	-
Information helpful	2.35	-	-	-
Information understandable	2.26	-	-	-
Process understood	-	207	37	-
Information received	-	166	65	-

* For the questions about program effectiveness, the questions about program personnel, and the first three questions about program components, respondents were given six choices that were rated on a scale of 1 (strongly agree), 2 (agree), 3 (neutral), 4 (disagree), 5 (strongly disagree), and 6 (not applicable).

In response to questions about program personnel and program components, a larger number of people agreed than disagreed that the staff were helpful and that the information received was helpful. Written comments received from citizen survey respondents were similar to the comments mentioned above.

Enteric Diseases and Sewage

With the advent of GIS technology at HCGHD, staff members began monitoring for any association between communicable-disease cases and the proximity of those cases to home sewage systems. In 1997, elevated numbers of Giardia cases were reported in Hamilton County. The cases appeared to be grouping around collector line discharges. Further investigation by public health nurses and sanitarians, however, determined that the exposure source was outside Hamilton County. The GIS technology will continue to be used at HCGHD as a public health surveillance tool.

What's Next? The operation permit program has reduced pollution and disease conditions. Odors and stream clarity have improved in many neighbourhoods. HCGHD plans to continue advancing water quality improvements, improving existing programs, and adding new ones. For example, work has begun on a program that will index stream conditions so that children may play and explore their backyard creeks and not worry about the threat of disease. Some program initiatives will be changed. Instead of assessing home sewage systems and water quality by political jurisdiction, a new watershed-based approach will be used. "Permits Plus" - the new major building and development permitting system of Hamilton County and the City of Cincinnati - coordinates all permitting-agency activity and will allow assessment by watershed. This will be a big help in evaluations. This automated land and infrastructure management system stores all data on a parcel of land in a computer file. A Specific project then can be monitored from concept to completion and thereafter for maintenance support. The integrated approach will provide up-to-date information on land use and infrastructure changes in a watershed and allow HCGHD to assess the impacts on water quality.

Another proposed project includes inspections of semipublic sewage disposal systems. It is estimated that 250 to 300 semipublic sewage disposal systems serve small businesses in Hamilton County. The health commissioner will review a contract between OEPA and HCGHD that will provide HCGHD the authority to inspect the semipublic systems. An initiative to educate homeowners about the importance of correct HSDS maintenance and operation could be accomplished by working with local coalitions. Comments from citizen surveys could help determine impact on subwatersheds from malfunctioning sewage systems and also help identify neighbourhood environmental issues.

HCGHD could also begin educating zoning commissions and elected officials about the importance of considering home sewage systems a part of the local utility infrastructure. Citizens in these decision-making roles tend to default land use decisions in unsewered areas to local boards of health. Future zoning decisions need to take into consideration the watershed impacts of changes in land use and the sewage utilities chosen to serve that use.

No one imagined, 45 years ago, the impact that indiscriminate use of HSDSs, especially aeration systems, would have on the neighbourhoods and backyard streams of Hamilton County. Improper siting and disregard of maintenance needs has created a costly cleanup for many citizens today. The passage of a new sewage code and the implementation of a multifaceted operation permit program has slowed further water quality degradation and facilitated needed remediation efforts. Reaching this point has taken a lot of effort and community support. In the years to come, HCGHD will strive to meet citizen expectations.

9

Water Politics: Technocratic Cooperation and Lasting Security

Water, like religion and ideology, has the power to move millions of people. Since the very birth of human civilization, people have moved to settle close to it. People move when there is too little of it. People move when there is too much of it. People journey down it. People write, sing and dance about it. People fight over it. All people, everywhere and every day, need it. —Mikhail Gorbachev

The distribution of environmental resources as a potential contributor to conflict has been the subject of considerable research, and these linkages have dominated the post-Cold War interest in environmental security. Within this genre much attention has been given to water resources, owing to their vital importance for human survival.

The distribution of environmental resources may contribute to conflict, but recent scholarship has begun to focus on the potential of environmental threats in stimulating conflict resolution. Uniting around a common aversion to environmental threats, as well as confidence-building through environmental cooperation, potentially hold great appeal for policymakers who aim to engage in proactive problem-solving rather than in precise problem identification. What is most significant for government decisionmakers to consider is that even if a conflict is not environmental in nature, the remedy may well be achieved through environmental means. Environmental cooperation may offer pathways to confidence-building or peacebuilding, whether or not the conflict has environmental roots.

This chapter explores the potentiality of such instrumental cooperation in the case of South Asia where regional conflict between two nuclear neighbours, India and Pakistan, is predicated in a history of religious rivalries and post-colonial demarcation. Despite inveterate antagonism, the two countries have managed to cooperate over water resources of the Indus River. How was this riparian cooperation enabled? And can it be reconfigured to provide for lasting peace in the region?

Anatomy of the Indus Waters Treaty

The Indian subcontinent quite literally owes its name to the waters of one river—the Indus. Regional politics are closely tied to the river's history and how different societies have used its waters for livelihood and for consolidating power. Hindu nationalists frequently recount that the very essence of their faith, dating back to the writings of the Rigveda in the second millennium B.C.E., is linked to the flow of the Indus. The name itself is a Latinized version of Sindhu, which means river in ancient Sanskrit, and from which the word "Hindu" and its concomitant ethnoreligious identity emerged. The partition of the subcontinent by the British in 1947 gave all but the very upper headwaters of the Indus to the newly formed Muslim majority country of Pakistan. More significantly, the major tributaries of the Indus that provided irrigation water for the fertile and densely populated region of Punjab on both sides of the border were divided. This was a classic conflict situation between upstream and downstream riparians, exacerbated by a lack of trust and intense territorial animosity between the two sides. This led to a series of disputes related to the Indus and its tributaries. Both countries tried to settle the matter bilaterally several times after partition but no lasting agreement was reached until the World Bank got involved as a mediating entity.

The resulting agreement, known as the Indus Waters Treaty, took nine years to negotiate and was signed in 1960. It is a particularly remarkable treaty since both sides have otherwise had tremendous hostility for one another and have defied efforts at cooperation. It is therefore instructive to consider the development and history of the treaty in greater detail as a potential model for regional environmental cooperation. The treaty is often cited as a success story of international riparian engagement, as it has withstood major wars between the two signatories (in 1965 and 1971), several skirmishes over water distribution and derivative territorial concerns. The agreement is also heralded as a triumph for the World Bank, which played an instrumental role in its negotiation during the height of the Cold War. The World Bank's role in this region was particularly unusual because India was a vanguard of the Non-Aligned Movement and wanted to disavow any pressure from international institutions or Western nations. The initiator and technical adviser of the agreement was David Lilienthal, the former head of the United States' Tennessee

Valley Authority, who suggested that an engineering perspective could contribute to resolving this political stalemate. After a visit to India and Pakistan in 1951, he advised the two countries to divide the Indus Basin geographically. India would have unrestricted use of the three eastern rivers (the Ravi, Sutlej and Bias), while Pakistan would completely control the three western rivers (the Jhelum, Chenab and Indus). The World Bank played a significant role by providing mediation, support staff, funding and proposals for pushing negotiations forward. Under the leadership of President David Black, the World Bank was able to persuade the international community to contribute nearly $900 million for impoundment construction.

Nine years after Lilienthal's initial visit, both countries were finally convinced to sign the agreement. The Indus Waters Treaty obligated Pakistan to build a canal system, which, by utilising previously less-developed rivers, decreased Pakistan's dependence on the Indus tributaries the treaty gave to India. The treaty also charged India and Pakistan with exchanging information and establishing joint monitoring mechanisms of river flow to ensure enforcement. The key provisions of the agreement are as follows:

- An agreement that Pakistan would receive unrestricted use of the western rivers, which India would allow to flow unimpeded, with minor exceptions;
- Provisions for three dams, eight link canals, three barrages and 2,500 tube wells to be built in Pakistan;
- A ten-year transition period, from 1 April 1960 to 31 March 1970, during which time water would continue to be supplied to Pakistan according to a detailed schedule;
- A schedule for India to provide its fixed financial contribution of $62 million in ten annual installments during the transition period; and,
- Additional provisions for data exchange and future cooperation.

As is often the case with riparian agreements, the treaty also established the Permanent Indus Commission, made up of one commissioner of Indus Waters from each country. In the technocratic spirit of the agreement, these representatives are often engineers rather than politicians. The two commissioners meet annually in order to:

- Establish and promote cooperative arrangements for implementation of the treaty;
- Promote cooperation between India and Pakistan in the development of the waters of the Indus system;
- Examine and resolve by agreement any question that may arise between the two countries concerning interpretation or implementation of the treaty; and,
- Submit an annual report to the two governments.

Both countries have upheld the Indus Basin Commission's information-sharing responsibilities; data on new projects, the water level in rivers and the water discharge of rivers are routinely conveyed to the other parties. If conflicts rise to the level of a dispute, the Indus River Commission will agree to mediation or arbitration, and the World Bank will appoint a neutral expert who is acceptable to both countries to resolve the dispute. Remarkably, although India and Pakistan constructed and carried out this agreement amidst skirmishes, threats and full-scale war, and even during armed conflict, neither country sabotaged the other's water projects. One of the water negotiators for Pakistan has commented that the role of international institutions is vital in making this enterprise function:

Both the parties are under the obligation of the Indus Waters Treaty, which asked the signatories not to disrupt the functioning of the commission. Any hurdle in the working of the commission is challengeable under the treaty, the guarantor of which is the World Bank.

No projects allowed under the treaty's provision of "future cooperation" have been submitted since 1960, nor have any water quality issues. There have, however, been several other disputes that have arisen over the years. The first issues arose from Indian non-delivery of some waters during 1965 to 1966 that became questions of procedure and of the legality of commission decisions. Negotiators resolved that each commissioner acted as a government representative and had the authority to make legally binding decisions. Another dispute involving the design and construction of the Salal Dam on the Chenab River in Jammu, India was resolved by way of bilateral negotiations. As noted in a recent World Bank study of Pakistan's water policy, India and Pakistan advocate conflicting principles of management: "equitable utilisation" and "no appreciable harm," respectively. Both sides continue to foster misgivings about the treaty but accept it as the best option in a time of conflict. From the Indian perspective, the 75 percent allocation of water to Pakistan represented a fundamental violation of equitable utilisation. From the Pakistani perspective, the allocation of only 75 percent of the water when it possessed 90 percent of the irrigated land was a violation of the principle of no appreciable harm. As a mark of how leadership can achieve reconciliation despite high tensions, former Pakistani President Ayub Khan is quoted in the aforementioned study as saying, we have been able to get the best that was possible ... very often the best is the enemy of the good and in this case we have accepted the good after careful and realistic appreciation of our entire overall situation.... The basis of this agreement is realism and pragmatism.

As part of a study of the Tarbela and Mangla dams (the two Pakistani impoundments constructed as a result of the treaty), the World Commission on Dams concluded that:

The Indus Waters Treaty represents the only ongoing agreement between India and Pakistan that has not been disrupted by wars or periods of high tension. Cooperation that builds on this treaty could not only present opportunities for better water management between those two countries, but also serve as a model for water-sharing arrangements between India, Bangladesh and Nepal.

Beyond Technical Cooperation: Prospects for Instrumental Peace

Although the Indus Waters Treaty has been able to overcome some minor issues (such as the Salal Dam dispute, which was resolved in 1978 through a new treaty), it has not been able to facilitate the resolution of larger conflicts, like Kashmir. The prospects for using the agreement over riparian issues as a means of conflict resolution more broadly can be traced back to a statement by U.S. Assistant Secretary of State George McGhee, who pointed out in 1951 that, a settlement of the canal waters question would signify those basic reversals of policy by the Governments of both India and Pakistan without which there can be no political rapprochement. Thus, the canal waters question is not only a functional problem, but also a political one linked to the Kashmir dispute.

As reported in the World Bank archives on this case, the British Prime Minister Anthony Eden felt that if this linkage were not possible, the resolution of the waters dispute could at least reduce tension over Kashmir. Interestingly enough, at one time it was argued by Pakistani politicians that the urgency of territorial claims on Kashmir for Pakistan also had a hydrological component. In 1957, the Pakistani prime minister, Hussain Suhrwardy, stated publicly that, "There are as you know six rivers (in the Indus Basin). Most of them rise in Kashmir. One of the reasons why, therefore, that Kashmir is so important for us is this water, these waters which irrigate our lands." However, since then, the Pakistani government has de-linked the Kashmir dispute from the reconciliation over water allocation. Commenting for this research on the potential of using the treaty as a conduit for resolving the Kashmir conflict, the Pakistani government's senior spokesmen on foreign policy, Mohammed Sadiq, stated the following:

The Indus Waters Treaty has been an important document for the water issue between the two countries. It has also helped in a framework for the resolution of water disputes in the region. Pakistan is fully committed to the treaty in letter and spirit. As far as the Kashmir dispute, this is not a water issue. It relates to the inalienable rights of Kashmiri people to self-determination.

As early as 1951, the Indian government has argued adamantly that: "The Canal Waters dispute between India and Pakistan has nothing to do with the Kashmir issue; it started with and is confined to the irrigation systems of East and West Punjab." Yet this decision to de-link the two has been made consciously by politicians, despite the ecological reality that Kashmir does indeed lay strategically within the headwaters of the river systems. In fact the Indus flows right through the valley corridor that connects Indian and Pakistani-held Kashmir. One can thus consider the cooperative role of water in this case at two levels. First, as suggested in the aforementioned statement by George McGhee, the resolution of the water dispute was a necessary but perhaps not a sufficient condition for conflict resolution over Kashmir. Second, since that condition for water cooperation has been met, the communication and opportunities for trust-building provided by the treaty continue to act as a potential means of further cooperation at the level of political psychology. Therefore, the Indus Waters Treaty has become the strongest link of cooperation between the two sides and, in times of crisis, it is often referenced as the ultimate cord of engagement that might be cut.

The latter proposition was put to the test in December 2001 following the Kashmiri militants' attack on the Indian Parliament two months prior, when India threatened to unilaterally abrogate the Indus Waters Treaty. However, six months later, the Permanent Indus Commission, which was established as part of the treaty, still met for the thirty-seventh time in New Delhi and the agreement weathered the story yet again. On a technical level, the Indus Waters Treaty was tested again when both India and Pakistan considered new dam projects to meet rising energy, demands. India is undertaking the Baglihar Hydropower Project (BHP) on the Chenab River in India, 160 kilometres north of Jammu, under severe opposition from Pakistan. Apart from objecting to the project design of the BHP, Pakistan has expressed opposition to the Tulbul navigation project, the Sawalkote Hydroelectric Project and the Kishanganga Hydroelectric Project, all located in Jammu and Kashmir. The Baglihar dispute was taken to the World Bank, which appointed a neutral technical expert, Swiss engineer Raymond Lafitte, in August 2005 to make a binding decision on the case. Lafitte gave his ruling on the dispute in early 2007 and the matter was amicably settled, with both sides claiming victory.

So far, the Indus Waters Treaty has served its purpose in de-escalating tensions over riparian water and has provided a direct avenue for regular, if technical, dialogue between the countries. It has not led to greater peacebuilding between the two countries as some of the original motivators of the treaty may have hoped. However, these most recent dam projects in Kashmir raise some potential prospects for using the agreement more instrumentally in resolving the Kashmir dispute. Increasingly, Kashmiri politicians are arguing that since the status of the territory is uncertain and so many of the disputes are in Kashmiri territory, they should be part of the Indus Basin negotiations as well. Whether such integrative solutions to the conflict would be found through cooperation on water remains to be seen, and is largely a question of leadership. Even when all the ingredients of rational state behaviour are in place, the ultimate action is dependent on individual leaders.

The Indus Waters Treaty may also be relegated to a broad range of confidence-building measures that countries may develop during times of crisis. As Shaista Tabassum has argued, the treaty did initially help to build some measure of conciliation between the two countries and was also framed as a "conflict avoidance measure." Soon after the treaty was signed, both countries did agree to negotiate actively on Kashmir and six rounds of talks were held from 1962 to 1964. However, the talks failed because of territorial intransigence on both sides and the escalation of domestic political pressures. It may also be argued that the de-linkage of the substantive issues related to the Indus Waters Treaty and the development of Kashmir as a region might have provided an opening for dialogue which was not availed. India's dominance as a hegemonic power in the region also gave it much more negotiating power that was not effectively countered by international pressure. For efficacy in such asymmetric circumstances, it is also important to consider the regional dynamics of cooperation over water.

Regional South Asian Strategies

South Asia has a remarkable history of cooperation over water-related issues in both maritime and riparian areas. India is South Asia's major littoral state, and shares maritime borders with several other South Asian states; in contrast, none of the other states have maritime borders with each other. India has settled its maritime boundaries with several of its neighbours, signing twelve bilateral agreements, including nine agreements with the Maldives, two each with Sri Lanka, Indonesia and Thailand, and one with Myanmar, as well as three trilateral agreements with Sri Lanka and the Maldives, Indonesia and Thailand, and Myanmar and Thailand. Pakistan has also signed two bilateral agreements to settle its maritime disputes—one with Oman and the other with Iran. However, maritime disputes continue between India, Pakistan and Bangladesh.

In the case of Bangladesh and India, the problem is not the maritime boundary, which can be defined fairly easily, but rather competing sovereignty claims over the island of Talpati. Bangladesh has a concave coast, and maritime boundaries in such geographical structures require integrative solutions and are extremely difficult to draw. Nevertheless, if a comprehensive settlement is reached in such cases, environmental factors can play a pivotal role since they help link various issues such as economic development and security. For example, a joint conservation monitoring arrangement can allow both sides access to areas that would otherwise be off-limits and give both sides an opportunity to cooperate in reducing environmental degradation. In particular, states that are ecologically vulnerable to extreme climatic events, such as Bangladesh, are recognising that poor environmental planning in coastal areas can have devastating economic impacts. The old environment/economy tradeoff is becoming less relevant as environmental pressures begin to have direct economic impacts. Pakistan's maritime dispute with India over the Sir Creek region could conceivably provide an opportunity to forge such a link between economic development and environmental cooperation.

In addition to maritime dispute settlements, several important river-sharing treaties have also been concluded in South Asia. India has agreements with Nepal, Bangladesh and Pakistan over riparian issues that are likely to be expanded in the future. Nepal, a small landlocked neighbour of India, is the upper riparian on the Mahakali River, which flows from Nepal into India. After protracted negotiations, the two states agreed on a treaty for the river in 1996. The importance of water negotiations was highlighted by the fact that the Nepalese parliament passed the treaty with the required two-thirds majority, despite a serious political crisis in Nepal at the time. According to commentator Krishna Rajan:

The treaty attracted attention in a number of countries as an important indication of the ability of India and Nepal as multiparty democracies to reach an agreement on cooperation on water resources on the basis of equality, transparency and equitable sharing of costs and benefits.... It does offer a model for India and Nepal on how to reach important understandings despite the uncertainties of democratic politics and coalition governments.

Also in 1996, India and Bangladesh signed a treaty on India's construction of the Farakkha Barrage, a dam that diverts the flow of the Ganges River into the Hooghly River during the dry season to flush silt from the port of Calcutta. The negotiations were spread over two decades and, after overcoming a number of controversies, finally concluded in the form of a thirty-year Farakkha Barrage Treaty. Regional organisations are often an important mechanism in promoting multilateral peacebuilding efforts. South Asia, as an example, has the potential to engage in such a process through the South Asian Association for Regional Cooperation (SAARC), which was established in 1985. While bilateral dispute resolution is excluded from SAARC's mandate, there are numerous aspects of bilateral disputes, which can have multilateral, or even global, implications. For example, the Siachen dispute between India and Pakistan has prevented scientists from studying glacial recession, hydrological impacts and climate change that can potentially influence the entire region. Arguments can thus be made that many of the so-called bilateral disputes that involve ecological factors have a salient global purpose. Despite discouraging signs that both quantitative and qualitative environmental issues (scarcity and pollution, respectively) have historically been relatively low on the priority list of decisionmakers in the region, it is important to note the establishment of SAARC was preceded by the formation of a regional environmental organisation. At the initiative of the United Nations Development Program, the South Asian countries—including Afghanistan and Iran—came together in 1980 and established the South Asian Cooperative Environmental Program (SACEP). The stated goal of SACEP at the time of establishment was: to promote regional co-operation in South Asia in the field of environment—both natural and human—in the context of sustainable development and on issues of economic and social development which also impinge on the environment and vice versa; to support conservation and management of natural resources of the region and to work closely with all national, regional and international institutions, governmental and nongovernmental, as well as experts and groups engaged in such co-operation and conservation efforts.

In its early years, SACEP was able to establish a "Regional Seas" program that had the potential to bring forth the territorial contentions for potential resolution. The interactions at a regional level through SACEP may well have helped to establish SAARC, which has a broader mandate in its charter of regional cooperation, covering a wide range of activities from energy to tourism to environmental protection, as well. While such instances of regional cooperation are promising, the South Asian case on its own does not provide us with enough structural coherence to develop an effective strategy, for moving forward with potential paths to making water an instrumental means of peacebuilding. Understanding the limitations of the current frames of policy analysis within international relations and considering alternative mechanisms for peacebuilding are important if we are to move beyond the self-fulfilling prophecy that tends to de-link environmental factors from peacebuilding.

Exploring Functionality of Water in Peacebuilding

Political geographer Kathryn Furlong has noted that dominant theories in international relations and international organisations tend to have five key flaws: 1) a mis-theorisation of hegemonic influences at work; 2) undue pessimism regarding the propensity for multilateral cooperation; 3) an assumption that conflict and cooperation exist along a progressive continuum; 4) a tenet that conflict is restricted to state competition; and 5) a depoliticization of ecological conditions. The Indus Waters Treaty exemplifies these challenges, which need to be addressed by scholars and practitioners alike. Theories of international relations that emphasize interdependence through mediating institutions such as the World Bank or the United Nations are most likely to offer some cooperative mechanisms in such asymmetric cases. The key to analysing environmental cooperation as a potential pathway to peacemaking is to dispense with notions of linear causality and instead consider conflict de-escalation processes as nonlinear (not having a simple cause and effect relationship), often constituting a complex series of feedback loops. Positive exchanges and trust-building gestures are a consequence of realising common environmental threats. Often, a focus on common environmental harms (or aversions) is psychologically more successful in leading to cooperative outcomes than a focus on common benefits, which may lead to competitive behaviour over the distribution of the gains. Specific research in game theory and operations research on the potential for cooperation over water is empirically showing that there are clear behavioural responses that suggest that such cooperation is possible.

We also appear to have history on our side in this regard. An important historical study on water conflicts conducted by Oregon State University has noted that "the rate of cooperation overwhelms the incidence of acute conflict." In the last fifty years, only thirty-seven disputes involved violence, and thirty of those occurred between Israel and one of its neighbours. Outside of the Middle East, researchers found only five violent events, while 157 treaties were negotiated and signed. The total number of water-related events between nations also favours cooperation: the 1,228 cooperative events are more than twice the number of 507 conflict-related events. Of these events, 62 percent are verbal, and more than two-thirds of these were not official statements.

Realist scholars argue that cooperation on environmental issues among adversaries merely constitutes "low politics" and does not translate into larger resolutions over high-level national security concerns. In this view, environmental conservation would be at best a means of diplomatic maneuvering between mid-level bureaucrats, and at worst a tool for influential elites to pursue their own narrow interests. Such critics give examples of cooperation on water resources between adversarial states such as India and Pakistan or Jordan and Israel without this cooperation translating into broader reconciliation or peace. Thus, it is presumed by some scholars looking at large historical data sets that environmental issues are not important enough in world politics to play an instrumental role in conflict resolution. Meanwhile, recent research conducted by the International Peace Research Institute in Norway has tried to extricate some of the various geographical aspects of cooperation and conflict potential of riparian states using regression analyses. The basic conclusion of this study is that a shared river basin tends to accentuate conflict, but a shared river boundary as a border does not. However, such studies cannot provide the granularity of analysis required to understand how cooperative mechanisms might still operate in cases such as the Indus, where the principal cause of the overarching conflict is not water.

One of the earliest contributions to the study of environmental peacebuilding was Peter Haas' work in the context of the Mediterranean Action Plan. Haas focused on ways in which knowledge exchange promotes environmental cooperation through the formation of what he termed "epistemic communities," networks of professional experts who arrive at shared views on scientific policy questions. These networks often take the form of civil society groups—sometimes facilitated by development donors—that exchange information on environmental issues. There is also a growing commitment from donors to "bioregionalism," the notion that ecological management must be defined by natural delineations such as watersheds and biomes rather than by national or other borders. Numerous joint environmental commissions between jurisdictions and countries have taken root all over the world, at times with implicit or explicit confidence- or peacebuilding goals. This evolution has also played out at various international forums in which bioregionalism and common environmental sensitivities have sometimes transcended traditional notions of state sovereignty. An important role for such organisations is to improve an understanding of interconnections between distributive competitive issues of environmental scarcity with the mutual loss of deteriorated quality of the resource in the absence of cooperation. Through such a process it may be possible to move functionally towards using water as a means of peacebuilding in South Asia and beyond.

The Indus Basin agreement has often been heralded as a success story of riparian cooperation between warring states. The role of the World Bank as the mediating institution in resolving this dispute between India and Pakistan is often cited as a positive intervention that led to a win-win outcome for all sides in the dispute. Yet the cooperation between the two states on this technical matter has not catalysed the resolution of the overarching conflict over the Kashmir region, giving some credence to realist assumptions about environmental factors being "low politics." A closer examination of the cooperative arrangements reveals that the cooperation may still have played an important role in deescalating tensions during times of crisis. Consequently, it is possible to link such arrangements to larger narratives of conflict over territory that may be deemed "high politics." A more positive framing of the case might reveal that water resources in this context are so important that adversaries must show some semblance of cooperation over them, even when that does not spill over into broader peace. Furthermore, the use of environmental issues in building peace must be considered over longer time horizons and repeated interactions, premised empirically on the following conditions:

- Development of a joint information base on a common environmental threat;
- Recognition that cooperation is essential to alleviate that threat;
- A cognitive connection and trust-building from initial environmental cooperation;

- Continued interactions over time due to environmental necessity;
- Clarification of misunderstandings and de-escalation of related conflicts; and,
- Increased cooperation and resultant peacebuilding.

These pathways are also considered the most empirically observed mechanisms, following a collective review by policy analysts for the United Nations Environment Program. The likelihood of environmental resources being used instrumentally in conflict resolution has increased in recent years. Certain environmental resources are now better understood as fundamental to basic economic, environmental and social processes, including sustaining human life. There is a growing realisation that environmental issues require integrated solutions across national borders since natural ecosystems do not recognise political boundaries. At the same time, politicians need to acknowledge that natural resources, particularly those as essential as water, can provide an important tool for resolving territorial disputes as well as providing a conduit for confidence-building measures between adversaries. Cooperation over water and the environment is also a potential way of avoiding conflict if we can frame the matter appropriately While South Asia has exemplified some parts of this framing routine, there is far more which can be accomplished if leaders are more willing to explore inherent ecological linkages between technical collaboration on water and lasting territorial security.

10

Ecological Sanitation

Ecological sanitation, also known as ecosan or eco-san, are terms coined to describe a form of sanitation that usually involves urine diversion and the recycling of water and nutrients contained within human wastes back into the local environment. An ecological sanitation (ecosan) viewpoint sees human waste and wastewater as an opportunity. When properly designed and operated, ecosan systems provide a hygienically safe, economical, and closed-loop system to convert human wastes into nutrients to be returned to the soil, and water to be returned to the land. Alternatively, solid wastes are converted into a biofuel. The primary application for ecosan systems has been in rural areas where connection to a sanitary sewer system is not possible, or where water supplies are very limited.

The main objectives of ecological sanitation are:

- To reduce the health risks related to sanitation, contaminated water and waste
- To prevent the pollution of surface and ground water
- To reuse nutrients or energy contained within wastes.

Reuse-oriented Sanitation Approaches

The recovery and use of urine and feces has been practiced by almost all cultures. The reuse was not limited to agricultural production. The Romans, for example, were aware of the bleaching attribute of the ammonia within urine and used it to whiten clothing. The most widely known reuse in agriculture has occurred in China. Reportedly, the Chinese were aware of the benefits of using excreta in crop production before 500 B.C., enabling them to sustain more people at a higher density than any other system of agriculture. The value of "night soil" as a fertilizer was recognized with well-developed systems in place to enable the collection of excreta from cities and its transportation to fields. However, its use promoted disease to such an extent that in Chinese cuisine almost all vegetables are thoroughly cooked.

Elaborate systems were developed in urban centers of Yemen enabling the separation of urine and excreta even in multi-story buildings. Feces were collected from toilets via vertical drop shafts, while urine did not enter the shaft but passed instead along a channel leading through the wall to the outside where it evaporated. Here, feces were not used in agriculture but were dried and burnt as a biofuel. In Mexico and Peru, both the Aztec and Inca cultures collected human excreta for agricultural use. In Peru, the Incas had a high regard for excreta as a fertilizer, which was stored, dried and pulverized to be utilized when planting maize. In the Middle Ages, the use of excreta and greywater was the norm. European cities were rapidly urbanizing and sanitation was becoming an increasingly serious problem, whilst at the same time the cities themselves were becoming an increasingly important source of agricultural nutrients. The practice of using the nutrients in excreta and wastewater for agriculture therefore continued in Europe into the middle of the 19th Century. Farmers, recognizing the value of excreta, were eager to get these fertilizers to increase production and urban sanitation benefited.

The increasing number of research and demonstration projects for excreta reuse carried out in Sweden from the 1980s to the early 21st century aimed at developing hygienically safe closed loop sanitation systems. Similar lines of research began elsewhere, for example in Zimbabwe, in the Netherlands, Norway and Germany. These closed-loop sanitation systems became popular under the name "ecosan", "dewats", "desar", and other abbreviations. They placed their emphasis on the hygenisation of the contaminated flow streams, and shifted the concept from waste disposal to resource conservation and safe reuse.

Concepts of Ecological Sanitation

Ecological sanitation (Ecosan) is based on an overall view of material flows as part of an ecologically and economically sustainable wastewater management system tailored to the needs of the users and to the respective local conditions. It does not favour a specific sanitation technology, but is rather a new philosophy in handling substances that have so far been seen simply as wastewater and water-carried waste for disposal. According to Esrey et al. (2003) ecological sanitation can be defined as a system that:

- Prevents disease and promotes health
- Protects the environment and conserves water
- Recovers and recycles nutrients and organic matter

Ecosan offers a flexible framework, where centralised elements can be combined with decentralised ones, waterborne with dry sanitation, high-tech with low-tech, etc. By considering a much larger range of options, optimal and economic solutions can be developed for each particular situation. Thus, the most important advantages of ecological sanitation systems are:

- Improvement of health by minimising the introduction of pathogens from human excreta into the water cycle
- Promotion of safe, hygienic recovery and use of nutrients, organics, trace elements, water and energy
- Preservation of soil fertility
- Contribution to the conservation of resources through lower water consumption, substitution of mineral fertiliser and minimisation of water pollution
- Improvement of agricultural productivity and food security
- Preference for modular, decentralised partial-flow systems for more appropriate cost-efficient solutions adapted to the local situation
- Promotion of a holistic, interdisciplinary approach
- Material flow cycle instead of disposal of valuable resources

Technologies of Ecosan Systems

Determining ecosan systems as ecological sanitation is not easy, for it is not just one specific technology, but a new approach based on an ecosystem-oriented view of material flows. The following diagram gives an overview of the different collection, treatment and reuse possibilities of the five flow streams considered in ecological sanitation systems:

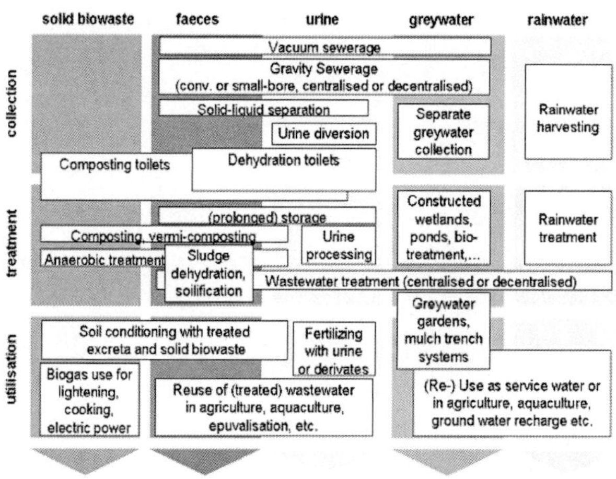

Figure: *Further information on ecosan technologies can be found in "Ecological Sanitation" by Winblad et al., in "Toilets that make compost" by Peter Morgan or in the gtz-ecosan technical data sheets, among other relevant literature.*

Project Examples

Examples of ecosan projects can be found among others in the collection of project data sheets of gtz ecosan or on the Enhanced Global Map of ecosan activities by EcoSanRes. In the following some examples are given that underline the diversity of ecosan projects:

Guangxi province, China - large-scale project of urine diverting dehydration toilets The dissemination programme of ecological dry toilets for Hsinchu County, Guangxi province, one of the poorest provinces in China, started in 1997 with support of UNICEF, SIDA and the Red Cross and has been expanded to 17 provinces until the year 2003. By this year, the scale of the project had increased to approximately 685,000 toilet units – today more than one million double vault urine diversion dehydration toilets (UDDTs) are installed in rural areas of China. In UDDTs, urine and faeces are collected separately: The urine is collected in the front and led by a plastic pipe to a storage canister from where it can be used as a fertilizer in agriculture, the faeces fall at the back in one of two ventilated storage chambers and are covered with ash for better dehydration. After about one year of storage the dried material can be removed and used as a soil conditioner in agriculture.

KfW, Frankfurt, Germany - vacuum toilets + greywater treatment The sanitation concept of the modern office building "Ostarkarde" of the KfW Bankengruppe in Frankfurt is based on a separate excreta and greywater collection. While urine and faeces are collected via vacuum toilets and a vacuum sewerage using much less water for flushing, the greywater from hand washing and kitchen is collected and treated separately in a compact activated sludge reactor combined with membrane filtration.

The treated greywater is then reused for toilet flushing and cleaning water. The amount of greywater can be reduced by 76% by this cost-efficient system which could be one of the prior choices for sanitation systems of newly constructed office buildings. Tanum Municipality in Sweden has introduced urine separation toilets to recover phosphorus.

Arguments for the Use of Ecological Sanitation

Often, water used in flush toilets is of drinking quality. Only 1% of global water is drinkable, therefore, it is a precious resource. Water fit to drink is being used for other purposes that can use lesser quality water, such as toilets.

Mixing feces and urine makes treatment difficult. All waste water treatment plants use some natural/biological processes, but nature does not normally have this waste water, so there are no microbes that can deal with this mix. In order to treat waste, treatment plants have to do this in stages. Each stage treats a different component of the mix by creating the right environment for microbes to do their work (aerobic, anaerobic, anoxic and the right pH). This is costly and requires energy. A mix of domestic and industrial effluent in water cannot be treated properly, for heavy metals and other pollutants make this water unsuitable for reuse. This is normally discharged into the ground or water bodies.

Because of the complexity of the treatment process, treatment plants tend to be large. This requires costly infrastructure to build and maintain it, often out of the reach of poorer communities. John Jeavons argues that "Each person's urine and manure contain approximately enough nutrients to produce enough food to feed that person." Urea is the major component of urine, yet we produce vast quantities of urea by using fossil fuels. By properly managing urine, treatment costs as well as fertilizer costs can be reduced. Feces also contains recognized nutrients, and could be used for modern agriculture, as micronutrient deficiency is a significant problem.

Environmental Design

Environmental design is the process of addressing surrounding environmental parameters when devising plans, programs, policies, buildings, or products. Classical prudent design may have always considered environmental factors; however, the environmental movement beginning in the 1940s has made the concept more explicit.

Environmental design can also refer to the applied arts and sciences dealing with creating the human-designed environment. These fields include architecture, geography, urban planning, landscape architecture, and interior design. Environmental design can also encompass interdisciplinary areas such as historical preservation and lighting design. In terms of a larger scope, environmental design has implications for the industrial design of products: innovative automobiles, wind-electricity generators, solar-electric equipment, and other kinds of equipment could serve as examples. Currently, the term has expanded to apply to ecological and sustainability issues.

History

The first traceable concepts of environmental designs focused primarily on solar heating, which began in Ancient Greece around 500 BCE. At the time, most of Greece had exhausted its supply of wood for fuel, leading architects to design houses that would capture the solar energy of the sun. The Greeks understood that the position of the sun varies throughout the year. For a latitude of 40 degrees in summer the sun is high in the south, at an angle of 70 degrees at the zenith, while in winter, the sun travels a lower trajectory, with a zenith of 26 degrees. Greek houses were built with south-facing faηades which received little to no sun in the summer but would receive full sun in the winter, warming the house. Additionally, the southern orientation also protected the house from the colder northern winds. This clever arrangement of buildings influenced the use of the grid pattern of ancient cities. With the North-South orientation of the houses, the streets of Greek cities mainly ran East-West. The practice of solar architecture continued with the Romans, who similarly had deforested much of their native Italian Peninsula by the first century BCE. The Roman *heliocaminus*, literally 'solar furnace', functioned with the same aspects of the earlier Greek houses. The numerous public baths were oriented to the south. Roman architects added glass to windows to allow for the passage of light and to conserve interior heat as it could not escape. The Romans also used greenhouses to grow crops all year long and to cultivate the exotic plants coming from the far corners of the Empire. Pliny the Elder wrote of greenhouses that supplied the kitchen of the Emperor Tiberius during the year.

Along with the solar orientation of buildings and the use of glass as a solar heat collector, the ancients knew other ways of harnessing solar energy. The Greeks, Romans and Chinese developed curved mirrors that could concentrate the sun's rays on an object with enough intensity to make it burn in seconds. The solar reflectors were often made of silver, copper or polished brass. Early roots of modern environmental design began in the late 19th Century with writer/designer William Morris, who rejected the use of industrialized materials and processes in wallpaper, fabrics and books his studio produced. He and others, such as John Ruskin felt that the industrial revolution would lead to harm done to nature and workers.

The narrative of Phil Cousineau's documentary film *Ecological Design: Inventing the Future* asserts that in the decades after World War II, "The world was forced to confront the dark shadow of science and industry." From the middle of the twentieth century, thinkers like Buckminster Fuller have acted as catalysts for a broadening and deepening of the concerns of environmental designers. Nowadays, energy efficiency, appropriate technology, organic horticulture and agriculture, land restoration, New Urbanism, and ecologically sustainable energy and waste systems are recognized considerations or options and may each find application.

By integrating renewable energy sources such as solar photovoltaic, solar thermal, and even geothermal energy into structures, it is possible to create zero emission buildings, where energy consumption is self-generating and non-polluting. It is also possible to construct "energy-plus buildings" which generate more energy than they consume, and the excess could then be sold to the grid. In the United States, the LEED Green Building Rating System rates structures on their environmental sustainability.

Examples

Examples of the environmental design process include use of roadway noise computer models in design of noise barriers and use of roadway air dispersion models in analyzing and designing urban highways. Designers consciously working within this more recent framework of philosophy and practice seek a blending of nature and technology, regarding ecology as the basis for design. Some believe that strategies of conservation, stewardship, and regeneration can be applied at all levels of scale from the individual building to the community, with benefit to the human individual and local and planetary ecosystems.

Specific examples of large scale environmental design projects include:

- Boston Transportation Planning Review
- BART - Bay Area Rapid Transit System Daly City Turnback project and airport extension.
- Metropolitan Portland, Oregon light rail system

Environmental Engineering Law

Environmental engineering law is a profession that requires an expertise in both environmental engineering and law. This field includes professionals with both a legal and environmental engineering education. This dual educational requirement is typically satisfied through an ABET accredited degree in environmental engineering and an ABA accredited law degree. Likewise, this profession requires both licensure in professional environmental engineering and admittance to one bar.

Environmental engineering law is the professional application of law, science and engineering principles to improve the environment (air, water, and/or land resources), to provide healthy water, air, and land for human habitation and for other organisms, and to remediate polluted sites. Environmental engineering lawyers seek to promote the advancement of technical engineering knowledge in the legal profession and to enhance informed legal analysis of complex environmental matters.

Practice Areas

Environmental engineering law professionals offer a sound knowledge base in the fields of both environmental engineering and law to address complex environmental problems which demand both professional technical practice and legal expertise. Areas of practice are continually expanding, but frequently include complex land transactions, such as:

- Brownfields redevelopment
- Asbestos baseline survey and building revaluation due to forthcoming asbestos abatements
- Soil contamination assessment & remediation, the development of a remedial action workplan (RAWP) and engineering controls, including an environmental land use restriction (ELUR)

- Total maximum daily load (TMDL) nutrient loading studies (ex. for NPDES wastewater discharges) and regulatory negotiation of nutrients discharge limits from waste treatment plants, such as phosphorus and nitrogen.

Environmental Impact Assessment

An environmental impact assessment is an assessment of the possible positive or negative impact that a proposed project may have on the environment, together consisting of the natural, social and economic aspects. The purpose of the assessment is to ensure that decision makers consider the ensuing environmental impacts when deciding whether to proceed with a project. The International Association for Impact Assessment (IAIA) defines an environmental impact assessment as "the process of identifying, predicting, evaluating and mitigating the biophysical, social, and other relevant effects of development proposals prior to major decisions being taken and commitments made." EIAs are unique in that they do not require adherence to a predetermined environmental outcome, but rather they require decision-makers to account for environmental values in their decisions and to justify those decisions in light of detailed environmental studies and public comments on the potential environmental impacts of the proposal.

EIAs began to be used in the 1960s as part of a rational decision making process. It involved a technical evaluation that would lead to objective decision making. EIA was made legislation in the US in the National Environmental Policy Act (NEPA) 1969. It has since evolved as it has been used increasingly in many countries around the world. As per Jay *et al.*(2006), EIA as it is practiced today, is being used as a decision aiding tool rather than decision making tool. There is growing dissent on the use of EIA as its influence on development decisions is limited and there is a view it is falling short of its full potential.There is a need for stronger foundation of EIA practice through training for practitioners, guidance on EIA practice and continuing research.

EIAs have often been criticized for having too narrow spatial and temporal scope. At present no procedure has been specified for determining a system boundary for the assessment. The system boundary refers to 'the spatial and temporal boundary of the proposal's effects'. This boundary is determined by the applicant and the lead assessor, but in practice, almost all EIAs address the direct, on-site effects alone.

However, as well as direct effects, developments cause a multitude of indirect effects through consumption of goods and services, production of building materials and machinery, additional land use for activities of various manufacturing and industrial services, mining of resources etc. The indirect effects of developments are often an order of magnitude higher than the direct effects assessed by EIA. Large proposals such as airports or ship yards cause wide ranging national as well as international environmental effects, which should be taken into consideration during the decision-making process.

Broadening the scope of EIA can also benefit threatened species conservation. Instead of concentrating on the direct effects of a proposed project on its local environment some EIAs used a landscape approach which focused on much broader relationships between the entire population of a species in question. As a result, an alternative that would cause least amount of negative effects to the population of that species as a whole, rather than the local subpopulation, can be identified and recommended by EIA.

There are various methods available to carry out EIAs, some are industry specific and some general methods:

- Industrial products - Product environmental life cycle analysis (LCA) is used for identifying and measuring the impact on the environment of industrial products. These EIAs consider technological activities used for various stages of the product: extraction of raw material for the product and for ancillary materials and equipment, through the production and use of the product, right up to the disposal of the product, the ancillary equipment and material.
- Genetically modified plants - There are specific methods available to perform EIAs of genetically modified plants. Some of the methods are GMP-RAM, INOVA etc.
- Fuzzy Arithmetic - EIA methods need specific parameters and variables to be measured to estimate values of impact indicators. However many of the environment impact properties cannot be measured on a scale e.g. landscape quality, lifestyle quality, social acceptance etc. and moreover these indicators are very subjective. Thus to assess the impacts we may need to take the help of information from similar EIAs, expert criteria, sensitivity of affected population etc. To treat this information, which is generally inaccurate, systematically, fuzzy arithmetic and approximate reasoning methods can be utilised. This is called as a fuzzy logic approach.

At the end of the project, an EIA should be followed by an audit. An EIA audit evaluates the performance of an EIA by comparing actual impacts to those that were predicted. The main objective of these audits is to make future EIAs more valid and effective. The two main considerations are:

- scientific - to check the accuracy of predictions and explain errors.
- management- to assess the success of mitigation in reducing impacts.

Some people believe that audits be performed as a rigorous scientific testing of the null hypotheses. While some believe in a simpler approach where you compare what actually occurred against the predictions in the EIA document. After an EIA, the precautionary and polluter pays principles may be applied to prevent, limit, or require strict liability or insurance coverage to a project, based on its likely harms. Environmental impact assessments are sometimes controversial.

Application

Australia: The history of EIA in Australia could be linked to the enactment of the U.S. National Environment Policy Act (NEPA) in 1970, which made the preparation of environmental impact statements a requirement. In Australia, one might say that the EIA procedures were introduced at a State Level prior to that of the Commonwealth (Federal), with a majority of the states having divergent views to the Commonwealth. One of the pioneering states was New South Wales, whose State Pollution Control Commission issued EIA guidelines in 1974. At a Commonwealth (Federal) level, this was followed by passing of the Environment Protection (Impact of Proposals) Act in 1974.

The Environment Protection and Biodiversity Conservation Act 1999 (EPBC) superseded the Environment Protection (Impact of Proposals) Act 1974 and is the current central piece for EIA in Australia on a Commonwealth (Federal) level. An important point to note is that this Commonwealth Act does not affect the validity of the States and Territories environmental and development assessments and approvals; rather the EPBC runs as a parallel to the State/Territory Systems. Overlap between federal and state requirements is addressed via bilateral agreements or one off accredition of state processes, as provided for in the EPBC Act.

The Commonwealth Level

The EPBC Act provides a legal framework to protect and manage nationally and internationally important flora, fauna, ecological communities and heritage places-defined in the EPBC Act as matters of 'national environmental significance'. Following are the eight matters of 'national environmental significance' to which the EPBC ACT applies:

- World Heritage sites;
- National Heritage places;
- RAMSAR wetlands of international significance;
- Listed threatened species and ecological communities;
- Migratory species protected under international agreements;
- The Commonwealth marine environment;
- Nuclear actions (including uranium mining); and
- National Heritage.

In addition to this, the EPBC Act aims at providing a streamlined national assessment and approval process for activities. These activities could be by the Commonwealth, or its agents, anywhere in the world or activities on Commonwealth land; and activities that are listed as having a 'significant impact' on matters of 'national environment significance'.

The EPBC Act comes into play when a person (a 'proponent') wants an action (often called a 'proposal' or 'project') assessed for environmental impacts under the EPBC Act, he or she must refer the project to the Department of Environment, Water, Heritage and the Arts (Australia).

This 'referral' is then released to the public, as well as relevant state, territory and Commonwealth ministers, for comment on whether the project is likely to have a significant impact on matters of national environmental significance. The Department of Environment, Water, Heritage and the Arts assess the process and makes recommendation to the minister or the delegate for the feasibility.

The final discretion on the decision remains of the minister, which is not solely based on matters of 'national environmental significance' but also the consideration of social and economic impact of the project.

The Australian Government environment minister cannot intervene in a proposal if it has no significant impact on one of the eight matters of 'national environmental significance' despite the fact that there may be other undesirable environmental impacts. This is primarily due to the division of powers between the States and the Federal government and due to which the Australian Government environment minister cannot overturn a state decision.

There are strict civil and criminal penalties for the breach of EPBC Act. Depending on the kind of breach, civil penalty (maximum) may go up to $550,000 for an individual and $5.5 million for a body corporate, or for criminal penalty (maximum) of seven years imprisonment and/or penalty of $46,200.

The State and Territory Level

Australian Capital Territory (ACT): EIA in ACT is administered with the help of Part 4 of the Land (Planning and Environment) Act 1991 (Land Act) and Territory Plan (plan for land-use).

New South Wales (NSW)

In New South Wales, the Environment Planning Assessment Act 1979 (EPA) establishes three pathways for EIA. The first is under Part 3A of the Act where Major Projects require an environmental assessment. The second is under Part 4 of the Act dealing with development control. If a project does not require approval under Part 3A or Part 4 it is then potentially captured by the third pathway, Part 5 dealing with environment impact assessment.

Northern Territory (NT)

The EIA process in Northern Territory is chiefly administered under the Environmental Assessment Act (EEA). Although EEA is the primary tool for EIA in Northern Territory, there are further provisions for proposals in the Inquiries Act 1985 (NT).

Queensland (QLD)

There are four main EIA processes in Queensland. Firstly, under the Integrated Planning Act 1997 (IPA) for development projects other than mining. Secondly, under the Environmental Protection Act 1994 (EP Act) for some mining and petroleum activities. Thirdly, under the State Development and Public Works Organization Act 1971 (State Development Act) for 'significant projects'. Finally, Environment Protection and Biodiversity Conservation Act 1999 (Cth) for 'controlled actions'.

South Australia (SA)

The local governing tool for EIA in South Australia is the Development Act 1993. There are three levels of assessment possible under the Act in the form of an environment impact statement (EIS), a public environmental report (PER) or a Development Report (DR).

Tasmania (TAS)

In Tasmania, an integrated system of legislation is used to govern development and approval process, this system is a mixture of the Environmental Management and Pollution Control Act 1994 (EMPCA), Land Use Planning and Approvals Act 1993 (LUPAA), State Policies and Projects Act 1993 (SPPA), and Resource Management and Planning Appeals Tribunal Act 1993.

Victoria (VIC)

The EIA process in Victoria is intertwined with the Environment Effects Act 1978 and the Ministerial Guidelines for Assessment of Environmental Effects (made under the s. 10 of the EE Act).

Western Australia (WA)

The Environmental Protection Act 1986 (Part 4) provides the legislative framework for the EIA process in Western Australia. The EPA Act oversees the planning and development proposals and assesses their likely impacts on the environment.

Canada

This section does not cite any references or sources. Please help improve this section by adding citations to reliable sources. Unsourced material may be challenged and removed. *(November 2010)*

The *Canadian Environmental Assessment Act* (CEAA) is the legal basis for the federal environmental assessment (EA) process. CEAA came into force in 1995. Legislative amendments were introduced in 2001 and came into force on October 30, 2003. EA is defined as a planning tool to identify, understand, assess and mitigate, where possible, the environmental effects of a project. Under the CEAA, all federal government departments and agencies are required to undertake an EA for projects relating to a physical work and for any proposed physical activity listed in the Inclusion List Regulations where it exercises one or more of the following CEAA triggers:

- Proposes or undertakes a project
- Grants money or any other form of financial assistance to a project
- Grants an interest in the land to enable a project to be carried out
- Exercises a regulatory duty in relation to a project, such as issuing a permit or license that is included in the Law List Regulations.

If a federal government department or agency exercises one or more of the above-mentioned triggers, it becomes a Responsible Authority (RA) under CEAA. As an RA, the federal department or agency in question must ensure that an EA is carried out in accordance with the CEAA and must consider the EA findings before a decision is made that could allow the project to proceed.

China

The Environmental Impact Assessment Law (EIA Law) requires an environmental impact assessment to be completed prior to project construction. However, if a developer completely ignores this requirement and builds a project without submitting an environmental impact statement, the only penalty is that the environmental protection bureau (EPB) may require the developer to do a make-up environmental assessment. If the developer does not complete this make-up assessment within the designated time, only then is the EPB authorized to fine the developer. Even so, the possible fine is capped at a maximum of about US$25,000, a fraction of the overall cost of most major projects. The lack of more stringent enforcement mechanisms has resulted in a significant percentage of projects not completing legally required environmental impact assessments prior to construction.

China's State Environmental Protection Administration (SEPA) used the legislation to halt 30 projects in 2004, including three hydro-power plants under the Three Gorges Project Company. Although one month later (Note as a point of reference, that the typical EIA for a major project in the USA takes one to two years.), most of the 30 halted projects resumed their construction, reportedly having passed the environmental assessment, the fact that these key projects' construction was ever suspended was notable.

A joint investigation by SEPA and the Ministry of Land and Resources in 2004 showed that 30-40% of the mining construction projects went through the procedure of environment impact assessment as required, while in some areas only 6-7% did so. This partly explains why China has witnessed so many mining accidents in recent years.

SEPA alone cannot guarantee the full enforcement of environmental laws and regulations, observed Professor Wang Canfa, director of the centre to help environmental victims at China University of Political Science and Law. In fact, according to Wang, the rate of China's environmental laws and regulations that are actually enforced is estimated to be barely 10%.

Egypt

Environmental Impact Assessment (EIA) EIA is implemented in Egypt under the umbrella of the Ministry of state for environmental affairs. The Egyptian Environmental Affairs Agency (EEAA) is responsible for the EIA services.

In June 1997, the responsibility of Egypt's first full time Minister of State for Environmental Affairs was assigned as stated in the Presidential Decree no.275/1997. From thereon, the new ministry has focused, in close collaboration with the national and international development partners, on defining environmental policies, setting priorities and implementing initiatives within a context of sustainable development.

According to the Law 4/1994 for the Protection of the Environment, the Egyptian Environmental Affairs Agency (EEAA) was restructured with the new mandate to substitute the institution initially established in 1982. At the central level, EEAA represents the executive arm of the Ministry. The purpose of EIA is to ensure the protection and conservation of the environment and natural resources including human health aspects against uncontrolled development. The long-term objective is to ensure a sustainable economic development that meets present needs without compromising future generations ability to meet their own needs. EIA is an important tool in the integrated environmental management approach. EIA must be performed for new establishments or projects and for expansions or renovations of existing establishments according to the Law for the Environment.

EU

The European Union has established a mix of mandatory and discretionary procedures to assess environmental impacts. European Union Directive (85/337/EEC) on Environmental Impact Assessments (known as the *EIA Directive*) was first introduced in 1985 and was amended in 1997. The directive was amended again in 2003, following EU signature of the 1998 Aarhus Convention. In 2001, the issue was enlarged to the assessment of plans and programmes by the so called *Strategic Environmental Assessment (SEA) Directive* (2001/42/EC), which is now in force. Under the EU directive, an EIA must provide certain information to comply. There are seven key areas that are required:

1. Description of the project
    - o Description of actual project and site description
    - o Break the project down into its key components, i.e. construction, operations, decommissioning
    - o For each component list all of the sources of environmental disturbance
    - o For each component all the inputs and outputs must be listed, e.g., air pollution, noise, hydrology
2. Alternatives that have been considered
    - o Examine alternatives that have been considered
    - o Example: in a biomass power station, will the fuel be sourced locally or nationally?
3. Description of the environment
    - o List of all aspects of the environment that may be affected by the development
    - o Example: populations, fauna, flora, air, soil, water, humans, landscape, cultural heritage
    - o This section is best carried out with the help of local experts, e.g. the RSPB in the UK
4. Description of the significant effects on the environment
    - o The word significant is crucial here as the definition can vary
    - o 'Significant' needs to be defined
    - o The most frequent method used here is use of the Leopold matrix
    - o The matrix is a tool used in the systematic examination of potential interactions
    - o Example: in a windfarm development a significant impact may be collisions with birds
5. Mitigation
    - o This is where EIA is most useful
    - o Once section 4 has been completed it will be obvious where the impacts will be greatest
    - o Using this information ways to avoid negative impacts should be developed

- o Best working with the developer with this section as they know the project best
- o Using the windfarm example again construction could be out of bird nesting seasons

6. Non-technical summary (EIS)
   - o The EIA will be in the public domain and be used in the decision making process
   - o It is important that the information is available to the public
   - o This section is a summary that does not include jargon or complicated diagrams
   - o It should be understood by the informed lay-person
7. Lack of know-how/technical difficulties
   - o This section is to advise any areas of weakness in knowledge
   - o It can be used to focus areas of future research
   - o Some developers see the EIA as a starting block for poor environmental management

The Netherlands

EIA was implemented in Dutch legislation on September 1, 1987. The categories of projects that require an EIA are summarised in Dutch legislation, the Wet milieubeheer. The use of thresholds for activities makes sure that EIA is obligatory for those activities that may have considerable impacts on the environment.

For projects and plans that fit these criteria, a EIA report is required. The EIA report defines a.o. the proposed initiative, it makes clear the impact of that initiative on the environment and compares this with the impact of possible alternatives with less a negative impact.

Hong Kong

EIA in Hong Kong, since 1998, is regulated by the *Environmental Impact Assessment Ordinance 1997*. The original proposal to construct the Lok Ma Chau Spur Line overground across the Long Valley failed to get through EIA, and the KCR Corporation had to change its plan and build the railway underground. In April 2011, the EIA of the Hong Kong section of the Hong Kong-Zhuhai-Macau Bridge was found to have breached the ordinance, and was declared unlawful. The appeal by the government was allowed in September 2011. However, it was estimated that this EIA court case had increased the construction cost of the Hong Kong section of the bridge by HK$6.5 billion in money-of-the-day prices.

India

The Ministry of Environment and Forests (MoEF) of India have been in a great effort in Environmental Impact Assessment in India. The main laws in nation are Water Act(1974), The Indian Wildlife (Protection) Act (1972), The Air (Prevention and Control of Pollution) Act (1981) and The Environment (Protection) Act (1986). The responsible body for this is Central Pollution Control Board. Environmental Impact Assessment (EIA) studies need a significant amount of primary and secondary environmental data. The primary data are those which need to be collected in the field to define the status of environment (like air quality data, water quality data etc.). The secondary data are those data which have been collected over the years and can be used to understand the existing environmental scenario of the study area. The environmental impact assessment (EIA) studies are conducted over a short period of time and therefore the understanding the environmental trends based on few months of primary data has its own limitations. Ideally, the primary data has to be considered along with the secondary data for complete understanding of the existing environmental status of the area. In many EIA studies, the secondary data needs could be as high as 80% of the total data requirement. EIC is the repository of one stop secondary data source for environmental impact assessment in India.

The Environmental Impact Assessment (EIA) experience in India indicates that the lack of timely availability of reliable and authentic environmental data has been a major bottle neck in achieving the full benefits of EIA. The environment being a multi-disciplinary subject, a multitude of agencies is involved in collection of environmental data. However, there is no single organization in India which tracks the data available amongst these agencies and makes it available in one place, in a form and manner required by practitioners in the field of environmental impact assessment in India. Further, the environmental data is not available in value added forms that can enhance the quality of the EIA. This in turn adversely affects the time and efforts required for conducting the environmental impact assessments (EIAs) by project proponents and also timely environmental clearances by the regulators. With this background, Environmental Information Centre (EIC) has been set up to serve as a professionally managed clearing house of environmental information that can be used by MoEF, project proponents, consultants, NGOs and other stakeholders involved in the process of environmental impact assessment in India. EIC caters to the need of creating and disseminating of organized environmental data for various developmental initiatives all over the country. EIC stores data in GIS format and makes it available to all environmental impact assessment studies and to EIA stakeholders in a cost effective and timely manner.

Malaysia

In Malaysia, Section 34A, Environmental Quality Act, 1974 requires developments that have significant impact to the environment are required to conduct the Environmental impact assessment.

Nepal

In Nepal, EIA has been integrated in major development projects since the early 1980s. In the planning history ofNepal, the sixth plan (1980–85), for the first time, recognized the need for EIA with the establishment of Environmental Impact Study Project (EISP) under the Department of Soil Conservation in 1982 to develop necessary instruments for integration of EIA in infrastructure development projects.

However, the government of Nepal enunciated environment conservation related policies in the seventh plan (NPC, 1985–1990). In order to enforce this policy and make necessary arrangements, a series of guidelines were developed, thereby incorporating the elements of environmental factors right from the project formulation stage of the development plans and projects and to avoid or minimize adverse effects on the ecological system.

In addition, it has also emphasized that EIAs of industry, tourism, water resources, transportation, urbanization, agriculture, forest and other developmental projects be conducted.

In Nepal, the government's Environmental Impact Assessment Guideline of 1993 inspired the enactment of the Environment Protection Act (EPA) of 1997 and the Environment Protection Rules (EPR) of 1997(EPA and EPR have been enforced since 24 and 26 June 1997 respectively in Nepal) to internalizing the environmental assessment system.

The process institutionalized the EIA process in development proposals and enactment, which makes the integration of IEE and EIA legally binding to the prescribed projects. The projects, requiring EIA or IEE, are included in Schedules 1 and 2 of the EPR, 1997 (GoN/MoLJPA 1997). Progresses were made in the Environmental protection issue during the 8th five year plan (1992–1997). The following development in Environmental protection were achieved during that time:

- Formulation of Environmental Protection Act 1997
- Establishment of Ministry of Environment
- Development of National Environmental Policies and Action Plan, EIA guidelines developed
- Consideration of environmental concerns in hydropower projects
- Development of industrial, irrigation and agricultural policies that undertook environmental concerns

New Zealand

In New Zealand, EIA is usually referred to as *Assessment of Environmental Effects* (AEE). The first use of EIA's dates back to a Cabinet minute passed in 1974 called Environmental Protection and Enhancement Procedures. This had no legal force and only related to the activities of government departments. When the Resource Management Act was passed in 1991, an EIA was required as part of a resource consent application. Section 88 of the Act specifies that the AEE must include "such detail as corresponds with the scale and significance of the effects that the activity may have on the environment". While there is no duty to consult any person when making a resource consent application (Sections 36A and Schedule 4), proof of consultation is almost certain to be required by local councils when making a decision about whether or not to publicly notifiy the consent application under Section 93.

Russian Federation

Russia holds the world's largest natural gas reserves, the second largest coal reserves, and the eighth largest oil reserves. Russia is also the world's largest exporter of natural gas, the second largest oil exporter and the third largest energy consumer.

As of 2004, the state authority responsible for conducting the State EIA in Russia has been split between two Federal bodies: 1) Federal service for monitoring the use of natural resources – a part of the Russian Ministry for Natural Resources and Environment and 2) Federal Service for Ecological, Technological and Nuclear Control . The two main pieces of environmental legislation in Russia are: The Federal Law 'On Ecological Expertise, 1995 and the 'Regulations on Assessment of Impact from Intended Business and Other Activity on Environment in the Russian Federation, 2000.

Federal Service for Monitoring the Use of Natural Resources

In 2006, the parliament committee on ecology in conjunction with the Ministry for Natural Resources and Environment, created a working group to prepare a number of amendments to existing legislation in order to cover such topics as stringent project documentation for building of potentially environmentally damaging objects as well as building of projects on the territory of protected areas. There has been some success in this area, as evidenced from abandonment of plans to construct a gas pipe-line through the only remaining habitat of the critically endangered Amur leopard in the Russian Far East.

Federal Service for Ecological, Technological and Nuclear Control

The government's decision to hand over control over several important procedures, including state EIA in the field of all types of energy projects, to the Federal Service for Ecological, Technological and Nuclear Control had caused a major controversy and criticism from environmental groups that blamed the government for giving nuclear power industry control over the state EIA. Not surprisingly the main problem concerning State EIA in Russia is the clear differentiation of jurisdiction between the two above-mentioned Federal bodies.

Sri Lanka

Environmental Impact Assessments: One popular approach to assist in smart growth in democratic countries is for law-makers to require prospective developers to prepare environmental impact assessments of their plans as a condition for state and/or local governments to go for Environmental Impact Assessments

These reports often indicate how significant impacts generated by the development will be mitigated - the cost of which is usually paid by the developer. These assessments are frequently controversial. Conservationists, neighborhood advocacy groups and NIMBYs are often skeptical about such impact reports, even when they are prepared by independent agencies and subsequently approved by the decision makers rather than the promoters. Conversely, developers will sometimes strongly resist being required to implement the mitigation measures required by the local government as they may be quite costly.

These assessments are frequently controversial. Conservationists, neighborhood advocacy groups and NIMBYs are often skeptical about such impact reports, even when they are prepared by independent agencies and subsequently approved by the decision makers rather than the promoters. Conversely, developers will sometimes strongly resist being required to implement the mitigation measures required by the local government as they may be quite costly.

The importance of the Environmental Impact Assessment as an effective tool for the purpose of integrating environmental considerations with development planning is highly recognized in Sri Lanka. The application of this technique is considered as a means of ensuring that the likely effects of new development projects on the environment are fully understood and taken into account before development is allowed to proceed. The importance of this management tool to foresee potential environmental impacts and problems caused by proposed projects and its use as a mean to make project more suitable to the environment are highly appreciated.

United States

Under United States environmental law an Environmental Assessment (EA) is compiled to determine the need for an *Environmental Impact Statement* (EIS), and originated in the National Environmental Policy Act (NEPA), enacted in 1969. Certain actions of federal agencies must be preceded by an EA or EIS. Contrary to a widespread misconception, NEPA does not prohibit the federal government or its licensees/permittees from harming the environment, nor does it specify any penalty if the EA or EIS turns out to be inaccurate, intentionally or otherwise. NEPA requires that plausible statements as to the prospective impacts be disclosed in advance. The purpose of NEPA process is to ensure that the decision maker is fully informed of the environmental aspects and consequences prior to making the final decision.

An *Environmental Assessment (EA)* is an environmental analysis prepared pursuant to the National Environmental Policy Act to determine whether a federal action would significantly affect the environment and thus require a more detailed *Environmental Impact Statement (EIS)*. The release of an Environmental Assessment results in either a *Finding of No Significant Impact (FONSI)* or an *Environmental Impact Statement (EIS)*.

The Environmental Assessment is a concise public document prepared by the federal action agency that serves to:

1. briefly provide sufficient evidence and analysis for determining whether to prepare an EIS or a Finding of No Significant Impact (FONSI)
2. Demonstrate compliance with the act when no EIS is required
3. facilitate the preparation of a EIS when a FONSI cannot be demonstrated

The Environmental Assessment includes a brief discussion of the purpose and need of the proposal and of its alternatives as required by CFR 102(2)(E), of the environmental impacts of the proposed actions and alternatives, as well as a listing of agencies and stakeholders consulted. The action agency must approve an EA before it is made available to the public. The EA is made public through notices of availability by local, state, or regional clearing houses, newspapers, etc. There is a 15-30 day review period required for an Environmental Assessment, while the document is made available for public commentary. An agency will release either a *Draft Environmental Assessment* (Draft EA) or a *Draft Environmental Impact Statement* (DEIS) for comment. Interested parties and the general public have the opportunity to comment on the draft, after which the agency will address all comments received and prepare a decision document, either a FONSI, Notice of Intent (NOI) to prepare a EIS or a Record of Decision for an EIS. The agency will then approve the " Final Environmental Assessment" (Final EA) or *Final Environmental Impact Statement* (FEIS). Commenting on the Draft EA is typically done in writing or email, submitted to the lead agency as defined in the Notice of Availability. Draft EIS's require public hearings, so comments can be made in person, as well as in writing. Occasionally, the agency will later release a "Supplemental Environmental Assessment " (Supplemental EA) or a *Supplemental Environmental Impact Statement* (SEIS), if the project parameters or environmental conditions change substantially after the issuance of the FONSI or ROD. The adequacy of an EIS can be challenged in federal court. Major proposed projects have been blocked because of an agency's failure to prepare an acceptable EIS.

One prominent example was the Westway landfill and highway development in and along the Hudson River in New York City. Another prominent case involved the Sierra Club suing the Nevada Department of Transportation over its denial of Sierra Club's request to issue a supplemental EIS addressing air emissions of particulate matter and hazardous air pollutants in the case of widening US Highway 95 through Las Vegas. The case reached the United States Court of Appeals for the Ninth Circuit, which led to construction on the highway being halted until the court's final decision. The case was settled prior to the court's final decision.

Several state governments that have adopted "little NEPAs," state laws imposing EIS requirements for particular state actions. Some those state laws such as ()the California Environmental Quality Act refer to the required environmental impact studies as environmental impact reports.

The structure of a generic Environmental Assessment is as follows:

1. Summary
2. Introduction
   - o Structure
   - o Background
   - o Purpose and Need for Action
   - o Proposed Action
   - o Decision Framework
   - o Public Involvement
   - o Issues
3. Alternatives, including the Proposed Action
   - o Alternatives
   - o Mitigation Common to All Alternatives
   - o Comparison of Alternatives
4. Environmental Consequences
5. Consultation and Coordination

These variety of state requirements are yielding voluminous data not just upon impacts of individual projects, but also to elucidate scientific areas that had not been sufficiently researched. For example, in a seemingly routine *Environmental Impact Report* for the city of Monterey, California, information came to light that led to the official federal endangered species listing of Hickman's potentilla, a rare coastal wildflower.

Transboundary Application

Environmental threats do not respect national borders. International pollution can have detrimental effects on the atmosphere, oceans, rivers, aquifers, farmland, the weather and biodiversity. Global climate change is transnational. Specific pollution threats include acid rain, radioactive contamination, debris in outer space, stratospheric ozone depletion and toxic oil spills. The Chernobyl disaster, precipitated by a nuclear accident on April 26, 1986, is a stark reminder of the devastating effects of transboundary nuclear pollution.

Environmental protection is inherently a cross-border issue and has led to the creation of transnational regulation via multilateral and bilateral treaties. The United Nations Conference on the Human Environment (UNCHE or Stockholm Conference) held in Stockholm in 1972 and the United Nations Conference on the Environment and Development (UNCED or Rio Summit, Rio Conference, or Earth Summit) held in Rio de Janeiro in 1992 were key in the creation of about 1,000 international instruments that include at least some provisions related to the environment and its protection.

The United Nations Economic Commission for Europe's Convention on Environmental Impact Assessment in a Transboundary Context was negotiated to provide an international legal framework for transboundary EIA. However, as there is no universal legislature or administration with a comprehensive mandate, most international treaties exist parallel to one another and are further developed without the benefit of consideration being given to potential conflicts with other agreements. There is also the issue of international enforcement. This has led to duplications and failures, in part due to an inability to enforce agreements. An example is the failure of many international fisheries regimes to restrict harvesting practises.

Annexed Projects

All projects are either classified as Annex 1 or Annex 2 projects. Those lying in Annex 1 are large scale developments such as motorways, chemical works, bridges, powerstations etc. These always require an EIA under the Environmental Impact Assessment Directive (85,337,EEC). Annex 2 projects are smaller in scale such as small installments, and maintenance works. These do not require an EIA.

Environmental Management

Environmental resource management is "a purposeful activity with the goal to maintain and improve the state of an environmental resource affected by human activities". It is not, as the phrase suggests, the management of the *environment* as such, but rather the management of the interaction and impact of human societies on the environment. Environmental resources management aims to ensure that ecosystem services are protected and maintained for equitable use by future human generations, and also, maintain ecosystem integrity as an end in itself by taking into consideration ethical, economic, and scientific (ecological) variables. Environmental resource management tries to identify the factors that have a stake in the conflicts that may rise between meeting the needs and protecting the resources.

Significance

Environmental resource management is an issue of increasing concern as reflected in its prevalence in seminal texts influencing global socio-political frameworks such as the Brundtland Commission's Our Common Future which brought to the fore the integrated nature of environment and international development and the Worldwatch Institute's annual State of the World (book series) reports.

Scope

Environmental resource management can be viewed from a variety of perspectives. Environmental resource management involves the management of all components of the biophysical environment, both living (biotic) and non-living (abiotic). This is due to the interconnected and network of relationships amongst all living species and their habitats. The environment also involves the relationships of the human environment, such as the social, cultural and economic environment with the biophysical environment.

Figure: *Improved agricultural practices such as these terraces in northwest Iowa can serve to preserve soil and improve water quality*

The essential aspects of environmental resource management are ethical, economical, social and technological which provide for formulation of principles and help in making decisions. The scientific and technical nature makes environmental resource management profession to operate in a humanistic and rational mode in the world.

Aspects

Ethical: Environmental resource management strategies are intrinsically driven by conceptions of human-nature relationships. Ethical aspects involve the cultural and social issues relating to the environment, and dealing with changes to it. "All human activities take place in the context of certain types of relationships between society and the bio-physical world (the rest of nature)", and so, there is a great significance in understanding the ethical values of different groups around the world. Broadly speaking, two schools of thought exist in environmental ethics: Anthropocentrism and Ecocentrism each influencing a broad spectrum of environmental resource management styles along a continuum.

These styles perceive "different evidence, imperatives, and problems, and prescribe different solutions, strategies, technologies, roles for economic sectors, culture, governments, and ethics, etc".

Anthropocentrism

Anthropocentrism, "an inclination to evaluate reality exclusively in terms of human values", is an ethic reflected in the major interpretations of Western religions and the dominant economic paradigms of the industrialised world. Anthropocentrism supports an understanding of nature as existing solely for the benefit of man and as a commodity to be used for the good of humanity and improved human quality of life. Anthropocentric environmental resource management is therefore not the conservation of the environment solely for the environment's sake, but rather the conservation of the environment, and ecosystem structure, for humankind's sake.

Ecocentrism

Ecocentrists believe in the intrinsic value of nature while maintaining an understanding that "human beings must use and even exploit nature to survive and live". It is this fine ethical line that ecocentrists navigate between "fair use and downright abuse". At an extreme end of the ethical scale, ecocentrism includes philosophies such as ecofeminism and deep ecology which evolved as a reaction to the dominant anthropocentric paradigms. "In its current form, it is an attempt to synthesize many old and some new philosophical attitudes about the relationship between nature and human activity, with particular emphasis on ethical, social, and spiritual aspects that have been downplayed in the dominant economic worldview".

Economic

The economy functions within, and is dependent upon goods and services provided by natural ecosystems. The role of the environment is recognized in both classical economics and neoclassical economics theories, yet the environment held a spot on the back-burner of economic policies from 1950 – 1980 due to emphasis from policy makers on economic growth. With the prevalence of environmental problems, many economists embraced the notion that "if environmental sustainability must coexist for economic sustainability, then the overall system must be one which permits the identification of an equilibrium between the environment and the economy". As such, economic policy makers began to incorporate the functions of the natural environment – or natural capital - particularly as a sink for wastes and for the provision of raw materials and amenities.

Figure: *A water harvesting system collects rainwater from the Rock of Gibraltar into pipes which lead to tanks excavated inside the rock.*

Debate continues among economists as to how to account for natural capital, specifically whether resources can be replaced through the use of knowledge and technology, or whether the economy is a closed system that cannot be replenished and is finite. Economic models influence environmental resource management in that management policies reflect beliefs about natural capital scarcity – if natural capital is believed to be infinite and easily substituted, environmental management would be irrelevant to the economy.

For example, economic paradigms based on neoclassical models of closed economic systems are primarily concerned with resource scarcity, and thus prescribe legalizing the environment as an economic externality for an environmental resource management strategy. This approach has often been termed 'Command-and-control'. Colby has identified trends in the development of economic paradigms, among them, a shift towards more ecological economics since the 1990's.

Ecological

"The pairing of significant uncertainty about the behaviour and response of ecological systems with urgent calls for near-term action constitutes a difficult reality, and a common lament" for many environmental resource managers. Scientific analysis of the environment deals with several dimensions of ecological uncertainty.

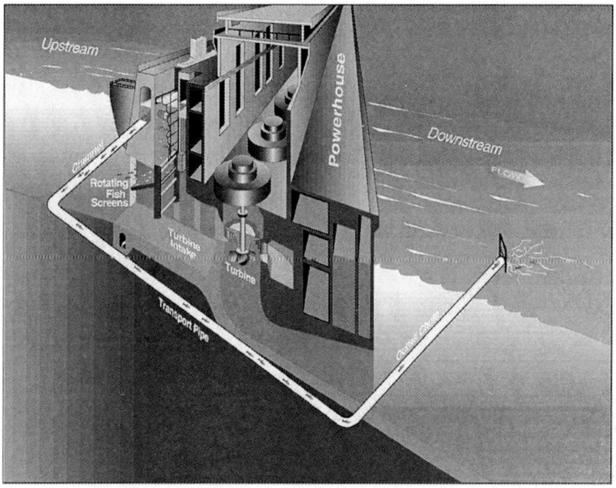

Figure: *A diagram showing the juvenile fish bypass system which allows young salmon and steelhead to safely pass the Rocky Reach Hydro Project in Washington*

Figure: *Fencing separates big game from vehicles along the Quebec Autoroute 73 in Canada.*

These include: *structural uncertainty* resulting from the misidentification, or lack of information pertaining to the relationships between ecological variables; *parameter uncertainty* referring to "uncertainty associated with parameter values that are not known precisely but can be assessed and reported in terms of the likelihood...of experiencing a defined range of outcomes"; and *stochastic uncertainty* stemming from chance or unrelated factors. Adaptive management is considered a useful framework through which to deal with situations of high levels of uncertainty though it is not without its detractors.

A common scientific concept and impetus behind environmental resource management is carrying capacity. Simply put, carrying capacity refers to the maximum number of organisms a particular resource can sustain. The concept of carrying capacity, whilst understood by many cultures over history, has its roots in Malthusian theory. An example is visible in the EU Water Framework Directive. However, "it is argued that Western scientific knowledge ... is often insufficient to deal with the full complexity of the interplay of variables in environmental resource management.

These concerns have been recently addressed by a shift in environmental resource management approaches to incorporate different knowledge systems including traditional knowledge, reflected in approaches such as adaptive co-management community-based natural resource management and transitions management among others.

Sustainability

Main article: Sustainability and environmental management

Sustainability and environmental resource management involves managing economic, social, and ecological systems within and external to an organizational entity in order for it to sustain itself and the system it exists within. In context, sustainability implies that rather than competing for endless growth on a finite planet, development will improve quality of life without necessarily having to consume more resources.

In order to sustainably manage the state of environmental resources affected by human activities organizational change is needed to instill sustainability values within an organization, in order to portray these values outwardly from all levels and to reinforce them in its surrounding stakeholder community. The end result should be a symbiotic relationship between the sustaining organization and community, along with the environment.

There are many drivers that compel environmental resource management to take sustainability issues into account. Today's economic paradigms do not protect the natural environment, yet they deepen human dependency on biodiversity and ecosystem services. Ecologically, massive environmental degradation and climate change threaten the stability of ecological systems that humanity depends on. Socially, an increasing gap between rich and poor and the global North-South divide denies many access to basic human needs, rights, and education, leading to further environmental destruction.

The planet's unstable condition is caused by many anthropogenic sources. As an exceptionally powerful contributing factor to social and environmental change, the modern organisation has the potential to apply environmental resource management with sustainability principals to achieve highly affective outcomes. To achieve sustainable development with environmental resource management an organisation should coincide with sustainability principals, such as: social and environmental accountability, long-term planning; a strong, shared vision; a holistic focus; devolved and consensus decision making; broad stakeholder engagement and justice; transparency measures; trust; and flexibility, to name a few.

Current Paradigm Shifts

In order to adjust to today's environment of quick social and ecological changes some organizations have begun to experiment with various new tools and concepts. Those which are more traditional and stick to hierarchal decision making are having difficulty dealing with the demand for lateral decision making that supports affective participation. Whether it be a matter of ethics or just strategic advantage organizations are internalizing sustainability principles. Examples of some of the world's largest and most profitable corporations who are shifting to sustainable environmental resource management are: Ford, Toyota, BMW, Honda, Shell, Du Pont, Swiss Re, Hewlett-Packard, and Unilever. An extensive study by the Boston Consulting Group reaching 1,560 business leaders from diverse regions, job positions, expertise in sustainability, industries, and sizes of organizations, revealed the many benefits of sustainable practice as well as its viability.

It is important to note that though sustainability of environmental resource management has improved, corporate sustainability, for one, has yet to reach the majority of global companies operating in the markets. The three major barriers to preventing organizations to shift towards sustainable practice with environmental resource management are: not understanding what sustainability is; having difficulty modeling an economically viable case for the switch; and having a flawed execution plan, or a lack there of. Therefore the most important part of shifting an organization to adopt sustainability in environmental resource management would be to create a shared vision and understanding of what sustainability is for that particular organization, and to clarify the business case.

Stakeholders

Public Sector: The public sector comprises the general government sector plus all public corporations including the central bank. In environmental resource management the public sector is responsible for administering natural resource management and implementing environmental protection legislation.

The traditional role of the public sector in environmental resource management is to provide professional judgement through skilled technicians on behalf of the public. With the increase of intractable environmental problems, the public sector has been led to examine alternative paradigms for managing environmental resources. This has resulted in the public sector working collaboratively with other sectors (including other governments, private and civil) to encourage sustainable natural resource management behaviours.

Figure: *A conservation project in North Carolina involving the search for bog turtles was conducted by United States Fish and Wildlife Service and the North Carolina Wildlife Resources Commission and its volunteers*

Private Sector

The private sector comprises private corporations and non-profit institutions serving households. The private sector's traditional role in environmental resource management is that of the recovers of natural resources. Such private sector recovery groups include mining (minerals and petroleum), forestry and fishery organisations. Environmental resource management undertaken by the private sectors varies dependent upon the resource type, that being renewable or non-renewable and private and common resources. Environmental managers from the private sector also need skills to manage collaboration within a dynamic social and political environment.

Civil Society

Civil society comprises associations in which societies voluntarily organise themselves into and which represent a wide range of interests and ties. These can include community-based organisations, indigenous peoples' organisations and non-government organisations (NGO). Functioning through strong public pressure, civil society can exercise their legal rights against the implementation of resource management plans, particularly land management plans. The aim of civil society in environmental resource management is to be included in the decision-making process by means of public participation. Public participation can be an effective strategy to invoke a sense of social responsibility of natural resources.

Tools

As with all management functions, effective management tools, standards and systems are required. An environmental management standard or system or protocol attempts to reduce environmental impact as measured by some objective criteria. The ISO 14001 standard is the most widely used standard for environmental risk management and is closely aligned to the European Eco-Management and Audit Scheme (EMAS). As a common auditing standard, the ISO 19011 standard explains how to combine this with quality management. Other environmental management systems (EMS) tend to be based on the ISO 14001 standard and many extend it in various ways:

- The Green Dragon Environmental Management Standard is a five level EMS designed for smaller organisations for whom ISO 14001 may be too onerous and for larger organisations who wish to implement ISO 14001 in a more manageable step-by-step approach,
- BS 8555 is a phased standard that can help smaller companies move to ISO 14001 in six manageable steps,
- The Natural Step focuses on basic sustainability criteria and helps focus engineering on reducing use of materials or energy use that is unsustainable in the long term,
- Natural Capitalism advises using accounting reform and a general biomimicry and industrial ecology approach to do the same thing,
- US Environmental Protection Agency has many further terms and standards that it defines as appropriate to large-scale EMS,
- The UN and World Bank has encouraged adopting a "natural capital" measurement and management framework,
- The European Union Eco-Management and Audit Scheme (EMAS).

Other strategies exist that rely on making simple distinctions rather than building top-down management "systems" using performance audits and full cost accounting. For instance, Ecological Intelligent Design divides products into consumables, service products or durables and unsaleables - toxic products that no one should buy, or in many cases, do not realize they are buying. By eliminating the unsaleables from the comprehensive outcome of any purchase, better environmental resource management is achieved without "systems". Recent successful cases have put forward the notion of "Integrated Management". It shares a wider approach and stresses out the importance of interdisciplinary assessment. It is an interesting notion that might not be adaptable to all cases.

Environmental Restoration

Environmental restoration is a term common in the citizens' environmental movement. Environmental restoration is closely allied with (or perhaps sometimes used interchangeably with) *ecological restoration* or *environmental remediation*. In the U.S., remediation is the term used more in the realms of industry, public policy, and the civil services.

In the 1987 edition of his book *Restoring the Earth: How Americans are Working to Renew our Damaged Environment*, scientific editor and writer John J. Berger defined environmental restoration (or "natural resource restoration") as follows: "*... A process in which a damaged resource is renewed. Biologically. Structurally. Functionally.*"

Bibliography

Anderson, Mary P. and William W. Woessner: *Applied Groundwater Modeling: Simulation of Flow and Advective Transport*. Academic Press, San Diego, 1992.

Asaithambi, S. : *Economics of Ground Water Management in India*, Abhijeet, Delhi, 2008.

Bhakar, S.R. : *Ground Water Hydrology : Theory and Practice*, Agrotech, Delhi, 2009.

Bhave, P.R. and R. Gupta: *Analysis of Water Distribution Networks*, Narosa, Delhi, 2011.

Biswas Asit K. : *Integrated Water Resources Management in South and South-East Asia*, , Oxford University Press, Delhi, 2001.

Bourne, Peter G.: *Water and Sanitation: Economic and Sociological Perspectives* Academic Press. Orlando. 1984.

Brooks, D.: *Water: Local-Level Management*, Ottawa, International Development Research Centre, 2002.

Cullis, A. : *Rainwater Harvesting: The Collection of Rainfall and Runoff in Rural Areas,* London, U.K.: IT Publications, 1986.

Erik Nissen-Peterson : *Rainwater Catchment Systems*, UK: Intermediate Technology Publications, 1999.

Feder, Paul A.: *A Guidebook to Groundwater: Resources and Education Opportunities in the Great Lakes Region*, Great Lakes Commission, Ann Arbor, 1993.

Fetter, Charles Willard: *Applied Hydrogeology*, Merrill Publishing Company, Columbus, OH. 1988.

Geraghty, James J.: *Water Atlas of the United States*, Port Washington, NY, 1973.

Gischler, Christiaan E.: *Water Resources in the Arab Middle East and North Africa*, Cambridge, Eng, Middle East & North African Studies Press, 1979.

Gurjar, Ram Kumar : *Geography of Water Resources*, Rawat, Delhi, 2008.

Hemenway, Toby : *Gaia's Garden: A Guide to Home-Scale Permaculture*, Vermont: Chelsea Green Publishing Company, 2000.

Jat, M.L. and S.R. Bhakar: *Ground Water Hydrology : Theory and Practice*, Agrotech, Delhi, 2009.

Kanmony, J. Cyril : *Drinking Water Management : Problems and Prospects*, Mittal Pub, Delhi, 2010.

LeRoy, K. : *Sustaining Water Washington*, D.C.: Population Services International, New York, 1993.

McGarry, M.G.: *Matching Water Supply Technology to the Needs and Resources of Developing Countries,* United Nations. New York. 1987

Meenu Bhatnagar: *Groundwater Management : Sustainable Approaches*, Icfai Books, Delhi, 2012.

Morris, Mary E.: *Water Scarcity and Security concerns in the Middle East*, Abu Dhabi, Emirates Center for Strategic Studies and Research, 1998.

Narwani, G.S. : *Community Water Management*, Rawat, Delhi, 2005.

Nielsen, David M.: *Practical Handbook of Groundwater Monitoring*. Lewis Publishers, Chelsea, 1991.

Palanisami, K : *Groundwater Management and Policies*, Macmillan Publishers India, Delhi, 2008.

Premjit Sharma: *Agricultural Drainage and Water Quality*, Gene Tech Books, Delhi, 2007.

Singh, A.K.: *Environment and Water Resources Management*, Adhyayan, Delhi, 2006.

Thomas, Harold E.: *The Conservation of Groundwater: A Survey of the Present Groundwater Situation in the U.S.,* McGraw Hill, New York, NY. 1951.

Todd, David Keith: *Groundwater Hydrology*. John Wiley & Sons, New York, NY. 1959.

William, K. : *Irrigation Investment Technology, and Management Strategies for Development Studies in Water Policy Management 9*, Boulder, Colo Westview Press, 1986.

Index

I

L

M

N

P

R

S

T

W

❑❑❑